Studying Mobile Media

Routledge Research in Cultural and Media Studies

For a full list of titles in this series, please visit www.routledge.com

Studying Mobile Media
Cultural Technologies, Mobile
Communication, and the iPhone

**Edited by
Larissa Hjorth, Jean Burgess,
and Ingrid Richardson**

Routledge
Taylor & Francis Group
NEW YORK   LONDON

First published 2012
by Routledge
711 Third Avenue, New York, NY 10017

Simultaneously published in the UK
by Routledge
2 Park Square, Milton Park, Abingdon, Oxon OX14 4RN

*Routledge is an imprint of the Taylor & Francis Group,
an informa business*

Typeset in Sabon by IBT Global.

Library of Congress Cataloging-in-Publication Data
Studying mobile media : cultural technologies, mobile communication,
    and the iPhone / edited by Larissa Hjorth, Jean Burgess, Ingrid Richardson.
        p. cm. — (Routledge research in cultural and media studies ; 39)
    Includes bibliographical references and index.
    1. Mobile communication systems—Social aspects.   2. Interpersonal
communication—Technological innovations—Social aspects.
3. iPhone (Smartphone)   4. Information technology—Social aspects.
5. Communication and culture—Technological innovations.   I. Hjorth,
Larissa.   II. Burgess, Jean (Jean Elizabeth)   III. Richardson,
Ingrid, 1966–
    HM1206.S78 2012
    004.16'7—dc23
    2011037026

ISBN13: 978-0-415-74839-1 (pbk)
ISBN13: 978-0-415-89534-7 (hbk)
ISBN13: 978-0-203-12771-1 (ebk)

First issued in paperback in 2013

Contents

1 Studying the Mobile
Locating the Field

Larissa Hjorth, Jean Burgess,
and Ingrid Richardson

INTRODUCTION

Over the past decade we have witnessed many transformations in the field of mobile communication as it moves unevenly into the smartphone era. This relatively short period has seen the emergence of a large body of literature that has addressed the multi-modal nature of, and need for interdisciplinary approaches to, mobile communication.[1] While much of this earlier work aimed to provide critical analyses of *communicative* practices surrounding mobile phone use, more recent texts have turned their attention to the mobile as an online, networked *media* device for games, video, music, and various other forms of everyday creativity. The smartphone's convergence of social, geolocative, and mobile media presents new challenges for how we study the increasingly mobile and interconnected fields of media production, distribution, and consumption. As an object of study, mobile media has expanded to encompass much more than just mobile communication—in turn attracting interest from internet studies, games studies, new media studies, and art, each discipline approaching mobile media devices from particular conceptual and methodological perspectives. So although mobile media scholarship has begun to grow,[2] it is by no means a coherent field of scholarship, and there are still many areas in need of elaboration and differentiation. What constitutes studying mobile media? For example, at what point is the study of mobile media the preoccupation of internet or game researchers? In its migration across a variety of areas, does the mobile media field have distinct characteristics? As the epitome of ubiquity, mobile media in an age of smartphones requires that we not only attempt to articulate the field but also to more systematically understand its various dimensions—the technical, cultural, social, political, and economic.

The iPhone represents a distinctive moment, both in the very short history of mobile media and in the much longer history of cultural technologies. Like the Walkman three decades earlier,[3] it marks a historical conjuncture in which notions about identity, individualism, lifestyle, and sociality—and their relationship to technology and media practice—require rearticulation. *Studying Mobile Media* explores not only the iPhone's particular characteristics, uses, and "affects," but also how the iPhone "moment" functions as a barometer for broader patterns of change, as well as the debates and controversies associated with them. By considering the iPhone "moment," *Studying Mobile Media* considers the number of convergent trajectories in the evolution of digital and mobile culture, and their implications for future scholarship. Through the lens of the iPhone—as a symbol, culture, and set of material practices around contemporary convergent mobile media, as well as a particular form of proprietary platform—the chapters included in this book explore some of the most productive available theoretical and methodological approaches for grasping media practice, consumer culture, and networked communication in the twenty-first century.

As a project, *Studying Mobile Media* took as a starting point Paul du Gay et al.'s "circuits of culture" approach, introduced in *Doing Cultural Studies*, which invited researchers to map the dynamics of culture—described by the core categories of consumption, production, regulation, representation and identity—as they co-influenced one another to produce the meanings of a particular cultural object (in that case, the Sony Walkman).[4] While the Sony Walkman case study itself is now decades old, the ongoing significance of this approach was captured by Gerard Goggin's more recent reinterpretation of it in *Cell Phone Culture*.[5] *Doing Cultural Studies* used an analysis of a significant technology of the time (the Sony Walkman) for dual purposes: to analyze new forms of media practice (consumption and production), and at the same time to consolidate and communicate the theoretical and methodological approaches of what was still an emerging discipline (Cultural Studies) for the benefit of students and researchers.

The ongoing relevance of *Doing Cultural Studies* beyond the "moment" of the Sony Walkman was that it captured the *zeitgeist* of cultural analysis in the 1990s and provided a benchmark for Cultural Studies research beyond that moment. *Studying Mobile Media* has similar aims: to use a single but highly complex cultural object (the iPhone) as a starting point, and to make available to readers a range of approaches and methods that may be useful in making sense of the contemporary proliferation of mobile media, social media, and user-created content; as well as contemporary trends in technological, cultural, and industrial convergence; and their implications for media, culture, and society. Through locating social mobile media within a diversity of cultural and industry contexts, *Studying Mobile Media* examines the ways in which technologies are both taken up and rejected by users. Indeed, *Studying Mobile Media* is as much interested in "discontinuous innovation" (products and applications that fail) as exploring the long-term effects of usage once the "honeymoon" period of new acquisition is over and the technology becomes integrated into users' everyday lives.

THE ARCHITECTURE OF THIS BOOK

Studying Mobile Media is not a simple celebration of the iPhone; rather, throughout the book the iPhone is deployed as a means of critically analyzing contemporary situated media practice—from social media and networked mobile media to creative industries such as the games industry. While the chapters do address the particularities of the iPhone—especially in terms of its photographic, game and geolocative media applications or "apps"—we also uncover some salient issues about the role of media in shaping, and being shaped by, locality, sociality and intimacy more broadly. Building upon the work conducted in mobile communication studies, *Studying Mobile Media* considers new forms of expression emerging from photographic applications, location-based services (LBSs), along with the associated phenomena of user-created content and social media, and the dynamics of convergence. The first section of the book provides both macro and micro perspectives on the iPhone as a cultural, technological and historical moment. Goggin's chapter provides an insightful analysis of the Apple Inc. machine in relation to the field of communication studies; Jean Burgess reimagines the open/closed platform debate in terms of the iPhone's position within the broader history of cultural technologies; the following two chapters offer cultural case studies of the iPhone as a moment in user practice: in China (Hjorth, Wilken, and Gu) and in Korea (Lee).

In Goggin's chapter, the iPhone is contextualized within broader mobile communication debates. Here, Goggin succinctly identifies not only how studying the iPhone can inform wider debates around mobile communication but also how the iPhone needs to be situated within such media ecologies. By addressing four key scholars in the area—James Katz, Christian Licoppe, Manuel Castells and Leopoldina Fortunati—Goggin provides a media compass with which to locate the iPhone. Contextualizing and situating the iPhone is also the agenda in Burgess's chapter; however, for Burgess, the iPhone needs to be understood as a moment in the longer and highly contested history of computer culture, and in particular, the way that user agency has been framed in the design and representation of personal computer technology. Burgess traces the shifts in Apple's corporate ethos, design and marketing practices from the late 1970s onward, arguing that earlier tensions between a "hacker" ethos and a market-driven populist one have resolved into a situation where highly usable but "closed" or "tethered" technologies—especially the iPhone—paradoxically provide the capacity for, rather than limiting, "cultural generativity."

Having situated the iPhone within two of its most important historical contexts, the book's next chapters mark the shift toward understanding the iPhone as a site of socio-cultural practice. Larissa Hjorth, Rowan Wilken, and Kay Gu consider the iPhone as a portal to geolocative media (geomedia) for young people in Shanghai, representing a particular "moment" within Chinese technoculture. Beyond the stories of Apple *shanzhai* (copy culture) stores, the iPhone presents a particular inroad into LBSs such as *Jie Pang*; this media practice is the preoccupation of a particular generation—in China they are called the *ba ling hou* (born between 1980 and 1989). As Hjorth et al. suggest, the deployment of *Jie Pang* through the iPhone reinforces culturally specific notions of social capital in the form of *guanxi.* From the Chinese context we migrate to Korea; in Dong-Hoo Lee's chapter, South Korean technoculture is identified as a dominant player in new media technologies through companies such as Samsung and LG. But as Lee notes, the introduction of the iPhone within the Korean context was quickly dubbed the "iPhone shock." In order to gain insight beyond the effects of this "shock," Lee firstly outlines the rise of mobile media communications in light of the smartphone spectacle and then moves onto a case study of Korean twenty-something iPhone users to gain a sense of the lived experience of mobile media practices in South Korea.

As a platform and phenomenon, we see how iPhone media practice—symbolized by its apps—provides numerous media cultures that both rehearse and extend practices across visual cultures, gaming and augmented reality. In "Part II: iPhone as a Platform and Phenomenon," we move through five very different and yet complementary case studies across photography, geomedia and games. In Palmer's, Chesher's and Verhoeff's discussions of iPhone photography, we see how apps both reinforce and depart from previous image practices. Camera phone practices amplify the local, highlighting the divergent ways in which public, private and the personal—especially converging around intimacy and intimate publics—are being reconfigured.[6] As Amparo Lasén and Edgar Gømez have observed in the case of the networked capacities of camera phones through online communities like Flickr, divides between public and private are undergoing significant change.[7] For Scott McQuire, the reconfiguration of public and private is part of the "new ways of conceptualizing the space and time of social experience and agency in a context in which the older boundaries of both territory and media are in a flux."[8]

These transformations are reflected in Palmer's discussion of the iPhone as part of the creative lifestyle and the questions this brings to bear on professional photography. In the first generation of camera phone studies, debates revolved around the erosion between amateur and professional users, especially concerning photo-journalism[9] and the way in which it interwove intimacy and co-presence[10] with banality.[11] The issue of a (networked) context, as it informed (vernacular) content, was key to these studies.[12] In light of the increased sophistication of the lens and editing suites, along with networked contexts such as Flickr,[13] Palmer considers whether one can talk about iPhone photography as a visual arts practice.

In Chesher's chapter, the context of iPhone photography is situated within a philosophical, rather than visual art, context. Moving beyond the first generation of camera phone studies that, according to Mizuko Ito and Daisuke Okabe,[14] was symbolized by the logic of ambient co-present intimacy in the form of three "s": sharing, storing and saving, Chesher argues that iPhone apps "transform, translate and transmit" images. Picking up on earlier discussions about the haunting of the analogue within digital photography (i.e., in programs such as Final Cut Pro, Adobe Photoshop), Chesher argues that iPhone apps further augment and simulate the nostalgic image. Drawing from Félix Guattari's essay "Machinic hetero-genesis," Chesher philosophically probes the limits and possibilities—in the form of specific Universes of reference—of iPhone photography espe-cially in terms of geomedia (location-based services). Nanna Verhoeff takes this philosophical analysis further by considering case studies of geome-dia and its impact upon situated and augmented reality. Unlike Chesher's chapter that focuses squarely on the iPhone, Verhoeff's discussion oscillates between the generalities of smartphone capabilities and the specificity of the iPhone, demonstrating the liminalities of the iPhone as a theoretical object. Drawing on structuralism, Verhoeff investigates the way in which iPhone navigation can be viewed as "a performative practice in mobile and interactive augmented reality tours."

Verhoeff's robust discussion of augmented reality within current mobile media practices provides a segue into the next chapter on mobile gaming and the iPhone. While the burgeoning of mobile games is not exclusive to the iPhone, the device does lend itself to particular kinds of game play and game cultures that are different from other handheld or haptic game con-soles.[15] In "Touching the Screen," Ingrid Richardson begins with a discus-sion of the micro- and macro-corporeal effects of touchscreen smart phones and the various embodiment relations specific to mobile gaming on such devices. She then turns to the particular affordances or "socio-somatics" of the iPhone as a game interface and offers a comparative analysis of loca-tion-based and casual iPhone games with respect to their distinct modali-ties of place, presence and being-in-the-world.

In the final section of the book, the discussion moves away from rep-resentation, consumption and practice toward the iPhone's conditions of production and its attendant forms of labor. Continuing the theme of iPhone games, John Banks turns our attention toward the specific model the iPhone's platform provides for industry development. By focusing upon iPhone games developer, HalfBrick, Banks considers the limits of innova-tion in relation to the iPhone as a development platform, and how this is both enabling and shaping practices within the games, software and content industries. In his vivid portrayal of the inhumane conditions surrounding iPhone production, Jack Linchuan Qiu considers the ugly side of the iPhone phenomenon. In "Network Labor" we are told about the worker suicides and the walled community of Foxconn, and finally left to reflect upon the cruel complicity involved in iPhone consumption. Qiu's moving depiction of Chinese workers in iPod City is followed by Hjorth's case study of female iPhone users in Australia in which she considers whether we can talk of an iPhone "affect." Consisting of both working mothers and non-mother users, Hjorth considers how the iPhone participates in the relationship between public and private, work and leisure boundaries. Through this case study, Hjorth asks whether the iPhone informs particular practices of personalization. Following on from Hjorth is Kate Crawford's "Four Ways of Listening with an iPhone," which discusses specific "vectors of listening" that traverse the iPhone and its users, along with other networked devices and media forms. Crawford identifies and critically examines a number of iPhone apps that demand particular kinds of individual and collective labor, including those that alter the way we engage with both sound and music in our immediate environment, allow us to perpetually listen-in to social networks and newsfeeds, enable "biometric listening" by monitoring one's mood, health and productivity, and finally "eavesdrop" on us by tracking and storing our location data. From the labors of listening, the final chapter of this collection provides us with a narrative of iPhone acquisition within an organization. Drawing from a case study of an art and design school, Ilpo Koskinen considers some of the paradoxes around professional creativities working within the university and the limits of the institution in affording space for those personalization practices. Universities, on the one hand, are big, often unwieldy organizations in which changes take much time and procedure. And yet, on the other hand, if a university prides itself on innovation—especially around new media technologies—it needs to be able to administer and implement new technologies quickly. Moreover, as universities increasingly become corporatized, their employees often have difficulties negotiating personal and work media.

Taking the iPhone both as a central object of study and as a departure point, *Studying Mobile Media* considers a range of socio-cultural and industrial factors informing the rise of new media practices within the context of networked and user-created content environments. In providing a critical snapshot of these factors and practices, *Studying Mobile Media* moves beyond the current dominance of bifurcated conceptual models that imagine users as either "empowered" or "exploited." By investigating the emerging creative practices associated with mobile media—practices that involve emotion, affect and sociality as well as new forms of storytelling and collaboration—*Studying Mobile Media* moves these debates forward. We hope that this consideration of the iPhone as a key technical and cultural "moment" will provide the impetus for deeper cross-disciplinary approaches to broader trends and trajectories within media culture, effectively going beyond existing studies of mobile media and communication.

ACKNOWLEDGEMENTS

We would like to thank the Australian Research Council's Cultural Research Network (CRN) for their support of an iPhone workshop in June 2009 at Queensland University of Technology. Also special thanks goes to CRN director, Professor Graeme Turner.

NOTES

1. James Katz, ed., *Handbook of Mobile Communication Studies* (Cambridge, MA: MIT Press, 2009); James Katz and Mark Aakhus, eds., *Perpetual Contact: Mobile Communication, Private Talk, Public Performance* (Cambridge: Cambridge University Press, 2002); Peter Glotz and Stefan Bertschi, eds.,

Thumb Culture: Social Trends and Mobile Phone Use (Bielefeld: Transcript Verlag, 2005); Manuel Castells, Mireia Fernández-Ardevol, Jack Linchuan Qiu and Araba Sey, *Mobile Communication and Society: A Global Perspective* (Cambridge, MA: MIT Press, 2007); Richard Ling and Peter Pedersen, eds., *Mobile Communication: Re-negotiation of the Social Sphere* (London: Springer-Verlag, 2005); Daniel Miller and Heather Horst, *Cell Phone* (Oxford and New York: Berg, 2005); Gerard Goggin, *Cell Phone Culture: Mobile Technology in Everyday Life* (London: Routledge, 2006); Mizuko Ito, Daisuke Okabe and Misa Matsuda, eds., *Personal, Portable, Pedestrian: Mobile Phones in Japanese Life* (Cambridge, MA: MIT Press, 2005); Richard Ling and Jonathan Donner, *Mobile Phones and Mobile Communication* (London: Polity Press, 2009).

2. Gerard Goggin and Larissa Hjorth, "Editorial: waiting to participate: emerging modes of digital storytelling, engagement and online communities," *Communication, Policy and Culture* 42(2), 2009: 1–5; Larissa Hjorth, *Mobile Media in the Asia-Pacific: Gender and the Art of being Mobile* (London: Routledge, 2009); Gerard Goggin, *Global Mobile Media* (London: Routledge, 2011).

3. Paul du Gay, Stuart Hall, Linda Janes, Hugh Mackay and Keith Negus, *Doing Cultural Studies: The Story of the Sony Walkman, Culture, Media and Identities* (London: Sage, 1997).

4. Ibid.

5. Goggin, 2006.

6. Larissa Hjorth, "Engagement rings: a cross-cultural analysis of camera phone genres, modes of sharing and digital storytelling," *The Future of Digital Media Culture: 7th International Digital Arts and Culture (DAC) Conference*, 15–18 September 2007, Perth, Australia http://www.beap.org/dac

7. Amparo Lasén and Edgar Gómez-Cruz, "Digital photography and picture sharing: redefining the public/private divide," *Knowledge, Technology and Politics* 22, 2009: 205–215.

8. Scott McQuire, *The Media City. Media, Architecture and Urban Space* (London: Sage, 2008), 20.

9. Gerard Goggin, "Calling the shots," *Sydney Morning Herald/Fairfax Digital*, 2 July 2005, http://smh.com.au/articles/2005/06/30/1119724747968. html (accessed 6 July 2005); Larissa Hjorth, "Snapshots," *Cultural Space and the Public Sphere in Asia*, hosted by Asia's Futures Initiative, 15–16 March 2006, Seoul; Larissa Hjorth, "Snapshots of almost contact: case study on South Korea," *Continuum* 21(2) 2007: 227–238.

10. Mizuko Ito and Daisuke Okabe, "Intimate Visual Co-Presence," presented at *UbiComp*, 2005 September 11–14, Takanawa Prince Hotel, Tokyo, Japan, http://www.itofisher.com/mito/ (accessed 10 December 2005).

11. Koskinen, 2007.

12. Ito and Okabe, op. cit.

13. Søren Mørk Petersen, "Common banality: the affective character of photo sharing, everyday life and produsage cultures," unpublished Ph.D. (Copenhagen, Denmark: IT University of Copenhagen, 2009).

14. Ito and Okabe, op. cit.

15. Larissa Hjorth and Ingrid Richardson, "The waiting game: complicating notions of (tele)presence and gendered distraction in casual mobile gaming," in Hajo Greif, Larissa Hjorth, Amparo Lasén and Claire Lobet-Maris, eds., *Cultures of Participation* (New York: Peter Lang: 2010), 111–125

Part I
iPhone as a Cultural Moment

2 The iPhone and Communication

Gerard Goggin

INTRODUCTION

Since its emergence in the late 1970s, the significance of the mobile phone for communication has been widely debated. People have been fascinated about how the mobile phone has changed human communication, what kinds of new communication practices have emerged, and what exactly the cultural implications of this technology might be. Not only has the mobile phone become an extraordinarily ubiquitous and versatile communication device—an emblem of our "liquid modernity"[1]; since the turn of the twenty-first century it has steadily moved center stage in shaping media—unrivaled only by the internet (with which it is increasingly in complex symbiosis). The questions first raised by the mobile phone in the 1980s and early to mid-1990s were those associated with the intriguing phenomenon of portable voice telephony—the emergence of a new, personal cultural technology, the way it confounded relationships between the private and public spheres and its distinctive take-up by particular groups in society, such as young people. With the phenomenon of text messaging, quickly followed by camera phones, mobile music and mobile games, the communication architecture of mobile devices became much more intricate—its potential for different kinds of communication all the richer. With the arrival of the iPhone in mid-2007, ushering in the era of the smartphone, there is a complex interplay unfolding: a layering of communication repertoires, a reshuffling of attributes and histories of old media and new concepts of the social function of mobiles—all challenging the adequacy of our theories of mobile communication.

To respond to this scenario, I start with a discussion of the work of four particular theorists—James Katz, Christian Licoppe, Manuel Castells and Leopoldina Fortunati. Each comes from a quite different tradition yet offers influential, rich ideas about how to think about mobile communication as its centrality to media grows. Drawing upon their ideas, I aim to provide a framework for thinking about the iPhone. In doing so, my argument is that an iPhone theory of communication needs to combine three key things derived from these theorists: a sense of the multi-modal, multimedia, multi-sensorial breadth of contemporary mobile communication; an account of the reconfiguring of older communication practices by newer ones in pursuit of the forging and managing of connections; and the power relations distinctive to iPhone communications.

COMMUNICATION THEORY:
MOBILE MEDIA AND THE DIGITAL AGE

In the pioneering collection *Perpetual Contact*, James E. Katz and Mark Aakhus offered an account of the mobile phone as global technology, which they termed a theory of *Apparatgeist* ("spirit of the machine").[2] Katz identified the wide range of uses of mobiles, especially as they have enlarged our sense of what media and communications are. In his 2011 volume *Mobile Communication: Dimensions of Social Policy*, Katz noted that the notion of perpetual contact still has relevance and indeed is growing:

> In terms of social grouping, the idea of near-perpetual contact is likely to continue to grow. Many new services are becoming available, and that will be available on popular mobile devices such as the Apple iPhone . . . While some sober and reasonable scholars will see this constant chattering, and tweeting, as deleterious to the inner reflection that makes life worthwhile, if not bizarrely unattractive to any reasonable human, it is also possible to see that such activities can actually form a bulwark against religious or political extremism. The value of multiple loyalties has been shown to be a strong counter-balance to totalitarian ideologies of all stripes . . . With mobile communication, a new way to have interstitial connection grows.[3]

Katz's interest in *connection* as a defining aspect of mobile communication has been discussed widely by other scholars.[4] One such scholar who has conducted especially interesting and pertinent research probing aspects of communication in mobile media is Christian Licoppe. In his classic 2004 paper on "connected presence," Licoppe argues the "technoscape" that we designate by terms such as the "information society" offers new resources, such as mobile phones, e-mail and messaging, that allow for the emergence of "connected" relationship management, "in which the (physically) absent party renders himself or herself present by multiplying mediated communication gestures up to the point where co-present interactions and mediated communication seem woven in a seamless web."[5] Licoppe's contention is that a "relationship is usually conducted over a variety of mediated interactions and that, to understand how a given relationship might be shaped by communication technologies, one needs to take into account the way the management of a given relationship will rely on the whole available technoscape."[6] In reviewing the social sciences literature on mediated interactions and relationships, Licoppe distinguishes three scales at which the problem can be addressed—the "micro" level appropriate to investigating single-mediated interactions, the "meso" level of social relationships and the "macro" level of social networks and communication fluxes. An integrated consideration—or at least awareness—of these three levels is important for Licoppe in following through on the conceptual insight that we dwell in the *cité connexionist* (or "connexionist world", as he translates the evocative phrase from Boltanski and Chiapello).[7] Such interaction among such levels is ultimately explained by a double-movement:

> From a historical point of view, we could consider that the management of relations in a "connected" mode is rooted in the current trend on the information technology scene. But these roots have the form of a global and mediated relationship economy in which even communication practices concerning the oldest communication devices are reshaped by the adoption of new ones.[8]

Here the example Licoppe proffers is the use of the landline telephone, in which calls in the pursuit of relationship maintenance and management are shortening in duration—something he sees as a paradoxical adaptation, moving us toward connected practices. Relationships are still vitally important in human life; however, with the telephone, and the mobile phone even more so, we see the potential—and indeed the rich significance—of mundane, brief calls, as well as long duration conversations, in communication. Of course, while noting the importance of connectedness and the new dimensions it entails, it is also critical to set out its conditions. Not only are there layers of connection, but also many disjunctures—not least from the frustrating, quotidian experiences of the difficulties making technologies work,[9] given the particular assemblages that they form, along with the social.[10]

The interaction among different scales of communication that we see in the work of Katz and explicitly theorized in Licoppe is something that has also been a feature of Manuel Castells' 2009 book *Communication Power*.[11] Castells' focus is in many ways on the macro level of communications network and flows. Yet, as with his previous work—especially the highly influential account of the "network society," as well as his co-authored work on global mobile communication[12]—Castells also proposes an account of how technology, the social, the role of the individual mind and interpersonal communication all need to be combined for an adequate theory of communication power. Especially germane for my purposes here, Castells stakes his theory upon the concept of "mass self-communication," which revises long-standing, classic models of how communication works. The kernel of Castells' theory is given in Chapter 2, "Communications in the Digital Age." Castells begins with a reprise of communication fundamentals, including the distinction between interpersonal communication and societal communication:

> In the former [interpersonal communication] the designated sender(s) and receiver(s) are the subjects of communication. In the latter, the content of communication has the potential to be diffused to society at large: this is what is usually called *mass communication*. Interpersonal communication is interactive . . . while mass communication can be interactive or one-directional . . . To be sure, some forms of interactivity can be accommodated in mass communication via other means of communication . . . Yet mass communication used to be predominantly one-directional.[13]

By contrast, Castells suggests a historically new form of communication has emerged with the diffusion of the internet that he terms "mass self-communication":

> It is mass communication because it can potentially reach a global audience, as in the posting of a video on YouTube, a blog with RSS links to a number of web sources, or a message to a massive e-mail list. At the same time, it is self-communication because the production of a message is self-generated, the definition of the potential receiver(s) is self-directed, and the retrieval of specific messages or content from the World Wide Web and electronic communication networks is self-selected.[14]

For Castells, the three forms of mass, self and mass self-communication co-exist and interact and result in something that is again unprecedented:

> What is historically novel, with considerable consequences for social organization and cultural change, is the articulation of all forms of communication into a composite, interactive, digital hypertext that includes, mixes, and recombines *in their diversity* the whole range of cultural expressions conveyed by human interaction [Castells' emphasis].[15]

For this new kind of communication to emerge, Castells suggests, there were distinct transformations that occurred in each dimension of the overall communication process. As well as the developments in technology, Castells nominates three other sites of profound change: the organizational and institutional structure of communication, which frame the definition and constitution of senders and receivers; the "cultural dimension" of the "multilayered transformation of communication," which he sees as unfolding via contradictory binaries of development of global culture side-by-side with multiple identity cultures, and individualism versus communalism; and the "expression of the social relationships, ultimately power relationships that underlie the evolution of the multimodal communication system."[16] Ultimately, in Castells' view, this is a conflictual, contradictory affair, fought out among powerful entrenched interests (especially those of the increasingly influential media, information and communications corporations), and social and individual actors wielding the new capacities of networked technologies: "The new field of communication in our time is emerging through a process of multidimensional change by conflicts rooted in the contradictory structure of interests and values that constitute society."[17]

After working through these dimensions, Castells returns to the sender-message-channel-receiver scheme of communication, a centerpiece in communications theory since the classic work of Shannon and Weaver. Castells' point of entry into these conversations is Umberto Eco's 1984 attempt in the paper "Does the Audience Have Bad Effects on Television?" to acknowledge the role of interpretation by receivers in communication, especially aggregated as audiences, by adding their own codes and sub-codes.[18] Castells places Eco's model of communication in the context of his own mass self-communication, adding the dimensions of multichannel and multimodal communication. He goes further, however, by proposing a fundamental change, one that arises from the fusion of sender and receiver. The implication of this inextricably merged dual role of sender and receiver is that a subject in the moment of reception needs to make sense of, and negotiate, the meaning as addressee of her own experience as sender.[19] For Castells, this revised scheme of communication is at the semiotic heart of what he sees as the "creative audience, the source of the remix culture that characterizes the world of mass self-communication."[20] Taking this further still, and drawing upon Henry Jenkins's account of convergence culture,[21] Castells is concerned to substantiate the hypothesis that:

> Convergence is fundamentally cultural and takes place, primarily, in the minds of the communicative subjects who integrate various modes and channels of communication in their practice and in their interaction with each other.[22]

Thus, Castells' grand theory offers one of the most comprehensive and detailed accounts of communication in recent times—and, especially in its emphasis on understanding creative audiences. Indeed this is an intriguing moment in the development of Castells' ideas, where in moving beyond his account of the "network society"—to theorize the kind of developments in the internet and mobiles since the mid-1990s especially—he seeks to bring together, in an integrated account, the different levels and modes of power. Yet there is much that remains to be elaborated further, especially in relation to mobile communication and the iPhone.

We need to bring in at this point the work of Leopoldina Fortunati, the scholar who has given us the most eloquent, elaborated and sustained social and political theoretical account of how power operates in the field of mobiles. Fortunati urges us to develop our accounts on mobile media into a "more complex paradigm which negotiates the new forms of labor and capital of postmodernity," especially grappling with the fact that "mobile media will undoubtedly intensify the gender and generational logics already at play [in the development of the mobile phone]."[23] While Castells talks of "social relationships" as "power relationships" that form the matrix of communication in the digital age, Fortunati offers a comprehensive account of how gender especially—intertwined with the dynamics of capital and labor—shapes both the substrate and the superstrate of mobile communication. In her theorizing of many cultural, social and political aspects of mobile technology—fashion, the human body, the family, work, news and journalism, social representations—Fortunati's work broadens our understanding of the many axes of power that shape communication, especially when it comes to the iPhone.

The works of Fortunati, Castells, Licoppe and Katz provide conceptual resources for us to think about the communication subfield of the iPhone and what is notable and distinctive about it. Katz's work draws our attention to the pervasiveness, breadth and variety of mobile communication. Katz also poses the question of connectedness, through the concept of perpetual contact and the global bearings of mobile technology. Licoppe's concept of "connected presence" combines a detailed investigation into the ethnography and pragmatics of facets of mobile communication with an awareness of the "technoscapes" and changing notions of the social which sub-tend it. To advance our understanding of this macro level, as well as offering a fresh account of the general economy of communication in the digital age, Castells' work is especially helpful. Finally, Fortunati broadens the vista of what logics of power are at play in mobile communication across bodies, identities, relationships, reproduction, labor and capital. This is crucial because of the new intensities and co-ordinates of cultural technology that the iPhone represents—as the life media and lifestyle device *par excellence* ("there's an app for everything").

So what relevance do these theories of communication have to the iPhone and its specific relation to communication? The short answer is that the iPhone challenges any easy summation or characterization of communication, mobile or otherwise. For one thing, the iPhone is the kind of intermingling of the internet *and* the mobile, to which Katz alludes. So our challenge is to understand how the iPhone is indeed more than the sum of its parts or of its bridging of mobile telecommunication, on the one hand, and the traditions of computing and the internet, on the other hand; not to mention bridging two still rather distinct bodies of medium-specific—or at least medium-skewed—research, represented by mobile communication studies and internet studies respectively. For another, the iPhone has assumed the guise of such a radically protean device—with applications, uses, meanings and practices that take it well beyond even the very recent but relatively established forms and functions of mobile and online media. Such metamorphoses in cultural technology that the iPhone augur require us to dig deep into our theoretical traditions for understanding communication.

TOWARD AN iPHONE THEORY OF
MOBILE COMMUNICATION

Apple has long had its own phenomenally successful *portable* media devices, in the form of iPods. With its genius for design, Apple's iPod was enormously popular in the digital music player market. While it communicative architecture was not about voice communications, nor text communications, it was certainly to do with the kinds of visual, aural, perceptual and other communications associated with music, design, fashion and technology.[24] By this, I mean that iPods were very much about how people signified their identities, comported themselves and otherwise communicated in specific contexts of social relations, across public and private boundaries. This is evident from advertising texts and other supports of the discourses of iPods, as well as the furious debates and highly invested analysis then engendered in discussions of connection, communities, relationships, publics and strangers.[25] What was becoming evident with the broader social and cultural shifts in which iPods were entailed was the complex sociality of communication and media consumption in such practices.[26] Mobile phones were becoming music devices, with their capacity to mimic and extend the capacities of portable music players, then digital music players. Mobiles also became the site for the invention of a new form of music—the ring tone.[27]

So if new assemblages of mobile phones and music players were emerging, then, something else was beginning to occur from the other side. The iPhone was Apple's spectacular foray into the world of mobiles. Like many other computing and software companies, Apple had long contemplated entering into the sphere of telecommunications. Apple's computing devices, notably its laptops, were increasingly being used as internet-connected devices, participating in the intensification of computer-mediated and net-worked-communication that Castells discusses in *Communication Power*. Apple had certainly prototyped cellular mobile phone devices previously, with its Rockr phone developed jointly with Motorola in an effort to devise a cell phone that drew on Apple's music capability.[28] However, it was with the development of its own iPhone that Apple was able to offer a full-fledged mobile phone that was widely taken up.

Apple did so by designing and developing its own mobile phone device and by forging new relationships with those companies controlling the telecommunications environment in which mobile communication typically operated. In the process, Apple reconfigured the power dynamics among network operators and vendors of handsets because it was able to conjure up, and indeed deliver upon, perceived demand for its product. Thus, it could strengthen its control as essentially a device manufacturer vis-à-vis network operators (gaining a greater share of revenue) and also vis-à-vis customers (neither permitting customers to modify a device in ways Apple did not wish, nor allowing them to easily port their device across different networks).

For at least these reasons, the iPhone held significant implications for accounts of communication power, in a relatively early stage of its introduction and adoption. The broad communicative aspects of user customization theorized by Larissa Hjorth as critically important and culturally significant aspects of gender and communication in the Asia-Pacific,[29] for instance, were pitched much more narrowly by Apple in its rather draconic rules on modification, where user reprogramming of the device could be deemed to have voided the warranty. Many users, of course, took great delight in gaily modifying, hacking and "jail-breaking" iPhones. Nonetheless, this dialectic between enlarging and constricting (or re-constraining) mobile communication is a theme in the research literature and user debates on the iPhone, as it develops.[30]

If we consider the iPhone from the perspective of theories of mobile communication, it could be suggested that many of the precepts of existing research might apply. The iPhone certainly allows voice communication to take place—acting in this sense as a classic telecommunications device. The iPhone is a telephone, in that it resembles a technology for receiving, initiating and conducting calls over distance. Further, it takes up the history of mobile telephony, through its design as a cell phone and use for mobile communication. The iPhone uses the cellular mobile telecommunications networks and is in this sense indistinguishable from other mobile phones. For instance, I can take my SIM card from one particular mobile phone—my favorite Nokia pre-smartphone model, or new Samsung Android smartphone—and insert this in an iPhone. The standards, protocols, regulatory and interconnection arrangements that make number portability and switching possible ensure that the iPhone will ring for my number, just as other phones will do. Thus, the iPhone reproduces the panoply of communicative attributes and features established through cycles of design and use over three decades of mobile phone use: protocols of mobile voice communication, building on those of telephone communication interaction;[31] address books and contact lists; clocks; alarms and notifications; text messaging; caller identification.[32] The iPhone also recapitulates and relies upon repertoires of embodied, tactic and affective practices, responses and association that grew to underpin mobile communication.[33]

So the iPhone has important similarities to other mobile phones and indeed builds upon the kinds of affordances we expect from this class of technology. Yet the iPhone also modifies these in a number of respects. The most immediately obvious change with which the iPhone was associated has to do with the senses. The first of the senses that Apple emphasized with its iPhone was touch. It presented a keyboard that needed to be visually perceived in order for the represented buttons to be pressed, but that did not have the tactile cues or anchoring of convention telephone or mobile phone keyboards. In addition, the iPhone allowed the use of touch—especially the sweep of a finger across its screen—to operate or launch applications, to turn pages or to increase or decrease the size of windows on display. This use of touch requires electric conductivity between finger and screen, a circuit that cannot be created by a gloved hand. Thus, the sense of touch—the field of haptics—became much more prominent in mobile communications.[34] Much of this we might see as bound up with haptic perception, but it is also clearly about communicating with, and through, touch. Thus, it is to do with haptic communication.

A further enlargement of mobile communication with the iPhone occurs through its incorporation of not just additional senses, but of sensors— that is, sensing technology. The most obvious of these is the accelerometer. Operating on three axes, the accelerometer allowed the iPhone screen to switch between the portrait and landscape viewing mode. Usually phone screens had utilized the portrait mode, with only some set to landscape (e.g., the Nokia Communicator, the iconic phone with its accompanying keyboard). However, the option of turning around your device and having the operating software re-orient the screen to match was certainly novel. It was especially useful for viewing photos or reading books, articles, internet pages or other texts on the landscape view of the iPhone—rather than being constrained to the narrow portrait view. As well as allowing the phone orientation feature to operate, something keenly marketed by Apple in its initial iPhone advertising, the accelerometer also greatly broadened the potential range of things a mobile phone could do. This emerged from the use of "apps," a second distinctive facet of iPhone mobile communication worth our consideration.

MOBILE APPS: FROM PHONE TO COMPUTER COMMUNICATION

As I have argued, we can certainly approach the iPhone from the standpoint of existing mobile phone devices, and the models of communication believed to spring from these. From this frame of reference, the iPhone can be seen to crystallize key developments in the expansion and transformation of what is regarded as mobile communication. It is certainly not revolutionary in technical invention, but Apple has a genius for collecting, curating and synthesizing various strands of new technology—with its capstone of consumer design and brand enthusiasm. Of all its innovations, the haptic stands out in capitalizing both on the possibilities of the technology and also the affective dimension of the staggeringly widespread incorporation of mobiles into our lives—the embodied cultures of mobile communication much analyzed in research, and richly attested to in all manner of everyday settings by users themselves. The haptic investment of the iPhone finds its way into its other great opening up of new pathways in mobile communication: apps.

While the features of the iPhone thus far discussed could be possible without downloadable mobile computing software, it is this teeming universe of rich, extensible metamorphoses that now gives us pause. Originally the iPhone launched as a device to which it was not possible to install software, without modifying the device—and only then with some difficulty. The arrival some months later of the App Store allowed software firms and inventors to develop applications for the iPhone. Once approved by Apple, these "apps," as they were called, were easily downloaded, stored, arranged, activated and updated by users. While data and software applications had been a much discussed and anticipated feature of mobile technologies since at least the late 1990s, the reality had been that these were relatively difficult for most users to install. With apps, an "arena for innovation" was created for developers. And the App Store also provided a simple way for users to discover, purchase and install software they liked.

This establishment of a platform and marketplace for software brought mobile computing alive, establishing it henceforth as a feature of mobile devices—in which Apple was in the vanguard but which other smartphones also quickly copied, contended with and sought to better. The significance of apps, as I have already prefigured, is not just that this software represented new things to do with mobile phone and mobile media devices. Rather, the apps, combined with the affordances of the iPhone, changed our perceptions of the devices themselves—and, as I shall argue, the scene of mobile communications. The software potentially allowed a metamorphosis of what was previously regarded as a phone—smart or otherwise. The famous "bowling ball" app turned the iPhone into a virtual bowling ball that could be brandished and swung, as with Nintendo Wii, by a player in a game. With a flashlight, or torch, application, the de facto widely improvised use of a cell phone as a way to illuminate a dark place in the middle of the night became a formal application in which the phone functioned to provide a ray of light. The "thing" that was the mobile phone—already often called the modern day "Swiss army knife" [35]—took on greatly broadened identities, although the novelty of this apps-enabled metamorphosis of mobile technology can be overstated in the first flush of love of the iPhone technology.[36] The apps phenomenon could easily be viewed in the *longue durée* of the development of the computer, whereby the design and development of software allowed capabilities, features and identities of technology to be possible that, while relying on particular kinds of hardware and architectures,[37] far exceeded those previously associated with them.[38] As well as this corrective to the app-istes and app-ism by placing these in a history of software, it prompts us also to consider the full-fledged movement of the mobile phone into the domain of computing.[39]

Like other smartphones, the iPhone calls up various histories of computers that have developed over centuries.[40] We might remind ourselves of the literatures that seek to recognize and theorize computer cultures and to understand these as new media.[41] With the iPhone, the question of communication previously routed through a consideration of mobile communication re-joins and updates discussions of computer communication.[42] This is no longer a matter of revisiting the concept of computer-mediated communication,[43] something that the rise of the internet and the scale, scope and quality of networked communications has so thoroughly revised. Rather, to conceive of an iPhone theory of communication that would of necessity represent the larger cultural moment and social function would be to grasp the distinct parallel histories and attributes of computers and data networks, on the one hand, crossing over with those of mobile telecommunications and personal technology devices, on the other. The name currently being given to this intermingling of technology and communicative architectures and potentialities is mobile internet.

SOCIAL COMMUNICATION: MEDIA ECOLOGIES OF MOBILE INTERNET

It is possible to see in my argument thus far a ceaseless expansion of the perimeters and potentialities of communication associated with the iPhone. Yet, it is also important to recognize the heightened role of the senses, emotions and affects activated in new ways by the iPhone. The iPhone enables new kinds of verbal, visual, tactile, affective and sensory communication. Such expanded modalities of communication need to be taken into consideration in relation to the common arguments that the new data and mobile practices made possible with such smartphones detracts from communication, whether via information overload, hyperconnectness or, on the contrary, types of withdrawing from engagement and participation in public society or even civil society or the private sphere. There are new intricacies that need to be explored about what the new politics of communication now are, along with the criss-crossing of such modalities—as is revealed by developments in mobile internet with the iPhone moment.

The iPhone has given a fillip to developments in mobile internet, which itself has grown fitfully since the late 1990s. The iPhone has provided a relatively easy-to-use interface for running the internet and its applications over mobile devices. The iPhone was not the first device to combine Wi-Fi and cellular mobile (and, of course, Bluetooth), but it certainly provided an easy facility to switch between these modes. What also became possible with the iPhone, once Apple permitted this to occur, was an integration—almost seamless—between voice and video over internet protocol applications. As well as Skype, a popular VoIP application at the present time is Viber, an app specifically designed for iPhone (and then developed for Android and BlackBerry also), promising free calls and better sound quality than a "regular call."[44] What is especially interesting about Viber for my purposes here is its apparently seamless bridging of the respective VoIP and cellular mobile habitus. The experience of receiving a Viber call is difficult to easily distinguish from an incoming cellular mobile call—both potentially have the same notification, depending on the ringtone or notification set. Further, Viber, like parasite to host, insinuates itself into the iPhone contacts list. When making a call from Viber, the app looks up the contact list, which in any case for those using mobile e-mail, may well automatically combine e-mail, Viber and phone contacts. Elsewhere Kate Crawford and myself have argued that the advent of mobile social media sees a qualitative change wrought in mobile technology, namely that we can no longer "look to the phone as a sealed, standalone and portable vessel of connection and engagement, but as a portal that opens into many other spaces."[45] The proliferation of varieties of telephony across internet, Bluetooth and cellular mobile networks is something that the iPhone catalyzes for the domestic market. In doing so, as is evident from this discussion of one of the iconic features of mobile communication—the contact list—the things bound up with such groups of people who might be contactable becomes rather more complex and ambiguous.

Another new ligature between mobiles and the internet assembled through the iPhone also harks back to an old theme in telephonic communication—video calling.[46] The FaceTime application on the iPhone 4 provided a useful fillip to videotelephony, though not as much as Apple might have thought: "People have been dreaming about video calling for decades. iPhone 4 makes it a reality."[47] Skype made video calling a reality, and applications like FaceTime are built on the back of this user acceptance. Indeed, how FaceTime works is a subject widely discussed by pundits after its launch. Essentially it relied upon users connecting via cellular phone service for the first call, then once Apple's server registers the phone number, it allows a connection to be made over a packet-switched (internet) connection. So a FaceTime video call can ultimately be made over a Wi-Fi connection.[48] We are still only at the early stages of understanding the kinds of communication that occur in mobile videotelephony. In one of the few studies available, Julian Morel and Licoppe observe that:

> Mobile video calls are also a particularly interesting mode of interaction in which the continuous production of images by both participants is woven into the fabric of interaction: since it is straightforward to orient the camera in any direction, frames are produced and inspected for their potential communicative intent, leading to particular problems in the management of visual contexts on the move.[49]

Thus, it is likely with the suturing of mobile video calls on the iPhone—knitting together both internet and cellular telephone modes—the visual dimension of mobiles will come into even sharper view as a problematic in communication.

The third aspect of communication highlighted in this conjuncture of mobile internet has to do with the iPhone catalyzing another nascent area—locative media. In April 2011 researchers Alasdair Allan and Pete Warden presented a paper to the *Where 2.0 Conference* at Santa Clara, California, highlighting the amount of location data routinely collected and stored on a typical iPhone 4 or third-generation (3G) iPad, occasioning widespread concern.[50] This resulted in a rising tide of understandable anxiety about the personal data required to underpin the widening array of location apps, especially those allowing users to interact with others in their locale such as *Foursquare*, Facebook Places, Google Latitude and other mobile social networking and navigation software. The affordances and politics of these "geomedia" apps,[51] especially as they figure on smartphones, are the subject of much current research and debate[52]—rendering even more complicated a general theory of iPhone communication. Space permits me to make only a preliminary observation that these emergent media ecologies of mobile internet certainly suggest, at the very least, that we need to make an imaginative leap to reconceive communication and the social. The cumulation of these vectors of communication only briefly discussed here underscores the intensive reassembling of the social the iPhone moment surely entails.

COMMUNICATION BEYOND THE iPHONE

Returning to my departure point: is it sensible to even think of a general theory of iPhone communication? Or, rather, an expanded theory of mobile communication, emboldened and provoked by the mobile multimedia,[53] computing and internet mash-up the iPhone and its smartphone conjuncture represents? My answer would certainly be in the affirmative, though I am not sure what choice we might otherwise have as communication theorists. Indeed I am reminded of the dilemma faced by an earlier generation of communication theorists in the middle of the twentieth century, as they sought to grasp the social and political implications of the new media system they sought to understand as mass communications.[54]

As I have outlined, the iPhone is a *combinatoire* of many previous modes of communications and, in the parlance of Marshall McLuhan, literally invokes and reconceives their extensions of human senses.[55] It is not so much the iPhone itself that casts this magic spell, rather, to adapt the phrase of Katz, it harnesses the magic already in the air[56] of mobile and wireless technologies. The kinds of connection enabled by the device, its network of actors, including users and things, we are still tracing out. Crucially, we face the difficult task of grasping the new kinds of power relations that frame the scene of iPhone communication, as a messy media ecology of mobile internet unfolds. The way Castells approaches the overarching constitution of such communication power is by drawing a distinction between "mass communication" and "mass self-communication," contending that the latter characterizes our age and, as such, represents a paradigm shift. While I am not convinced that this two-fold distinction is ultimately tenable, what Castells does—as do Katz, Licoppe, Fortunati, as well as Hjorth, Richardson, and Wilken, and other theorists also—is draw our attention to the need to grasp the rich opportunities and complexities of the convergence of multiple modes of communication in mobile media, that perforce applies to the iPhone.

Such critical concepts and accounts of communication are vitally important in negotiating the potential and pitfalls of mobiles as cultural technologies in the age of the smartphone. The emergent infrastructure of iPhones, smartphones and "apps" potentially constitutes a new platform for the work of culture, for social action, for political participation and for supporting new economic models. Yet mobile communication, like other kinds of communication, can be an apparatus for control, framed by its own distinctive ideology. This is certainly evident in the unresolved tension between the representation of the iPhone as all-purpose personal machines, (as portrayed in Apple's advertising) and the fervor with which the technology has been embraced by users, on the one hand, and, on the other hand, the tight controls and commodification of software, intellectual property, freedom of expression, and access to technology, its use, and modification that the enclosed world of apps and iPhones for which Apple has been stridently criticized.[57]

ACKNOWLEDGEMENTS

My thanks to Larissa Hjorth, Jean Burgess and Steven Maras for helpful comments on this paper.

NOTES

1. Zygmunt Bauman, *Liquid Modernity* (Cambridge: Polity Press, 2000).
2. James E. Katz and Mark Aakhus, "Conclusion: making meaning of mobiles—a theory of *Apparatgeist*," in James E. Katz and Mark Aakhus, eds., *Perpetual Contact: Mobile Communication, Private Talk, Public Performance* (Cambridge: Cambridge University Press, 2002), 301–318.
3. James E. Katz, "Conclusion," *Mobile Communication: Dimensions of Social Policy* (Piscataway, NJ : Transaction, 2011), 306.
4. On social connection and mobiles, see contributions to Rich Ling and Scott Campbell, eds., *Mobile Communication: Bringing Us Together and Tearing Us Apart* (Piscataway, NJ: Transaction, 2011).
5. Christian Licoppe, " 'Connected' presence: the emergence of a new repertoire for managing social relationships in a changing communication technoscape," *Environment and Planning D: Society and Space* 22 2004: 135.
6. Ibid.
7. Luc Boltanski and Ève Chiapello, *Le Nouvel Esprit du Capitalisme* (Paris: Galliamard, 1999); published in English as *The New Spirit of Capitalism*, trans. G. Elliot (London: Verso, 2005).
8. Licoppe, op. cit., 153.
9. On the messy complexities of technology, notably infrastructure, see Brian Larkin, *Signal and Noise: Media, Infrastructure and Urban Culture in Nigeria* (Durham, NC: Duke University Press, 2008).

10. Bruno Latour, *Reassembling the Social: An Introduction to Actor-Network-Theory* (Oxford: Clarendon, 2005).
11. Manuel Castells, *Communication Power* (Oxford and New York: Oxford University Press, 2009).
12. Manuel Castells, Mireia Fernández - Ardèvol, Araba Sey and Jack Linchuan Qiu, *Mobile Communication and Society: A Global Perspective* (Cambridge, MA: MIT Press, 2006).
13. Castells, *Communication Power*, 55.
14. Ibid., 55.
15. Ibid., 55.
16. Ibid., 57.
17. Ibid., 57.
18. Umberto Eco, "Does the audience have bad effects on television?" in Robert Lumley, ed., *Umberto Eco: Apocalypse Postponed* (Bloomington, IN: Indiana University Press, 1994), 87–102.
19. Castells, *Communication Power*, 132.
20. Ibid.
21. Henry Jenkins, *Convergence Culture: Where Old and New Media Collide* (New York: New York University Press, 2006).
22. Castells, *Communication Power*, 135.
23. Leopoldina Fortunati, "Gender and the Mobile Phone," in Gerard Goggin and Larissa Hjorth, eds., *Mobile Technology: From Telecommunications to Media* (New York: Routledge), 23–34.
24. Paul du Gay, Stuart Hall, Linda Janes, Hugh Mackay and Keith Negus, *Doing Cultural Studies: The Story of the Sony Walkman* (London and Thousand Oaks: Sage, 1997); Michael Bull, *Sounding Out the City: Personal Stereos and the Management of Everyday Life* (Oxford: Berg, 2000).
25. See for instance, Michael Bull, *Sound Moves: iPod Culture and Urban Experience* (London and New York: Routledge, 2007).
26. Kenton O'Hara and Barry Brown, eds., *Consuming Music Together: Social and Collaborative Aspects of Music Consumption Technologies* (Dordrecht: Springer, 2006).
27. Gerard Goggin, *Global Mobile Media* (London: Routledge, 2011).
28. Gerard Goggin, "Adapting the Mobile Phone: The iPhone and its Consumption," *Continuum*, 23.2, 2009: 231–244
29. Larissa Hjorth, *Mobile Media in the Asia Pacific: Gender and the Art of Being Mobile* (London and New York: Routledge, 2009). See also: Leopoldina Fortunati, "User design and the democratization of the mobile phone," *First Monday* 7 (2006) http://www.uic.edu/htbin/cgiwrap/bin/ojs/index.php/fm/article/view/1615/1530
30. Perhaps the first scholarly text to critique the iPhone for its lack of openness was Jonathan Zittrain in his *The Future of the Internet and How to Stop It* (New Haven, CT: Yale University Press, 2008). Cf. David L. Altheide, *An Ecology of Communication: Cultural Formats of Control* (New York: Aldine de Gruyter, 1995).
31. Robert Hopper, *Telephone Conversation* (Bloomington: Indiana University Press, 1992).
32. Eija-Liisa Kasesniemi, *Mobile Messages: Young People and a New Communication Culture* (Tampere: Tampere University Press, 2003); Rich Ling, *The Mobile Connection: The Cell Phone's Impact on Society* (San Francisco, CA: Morgan Kaufmann, 2004); James E. Katz, ed., *Handbook of Mobile Communication Studies* (Cambridge, MA: MIT Press, 2008); Leslie Haddon and Nicola Green, *Mobile Communications: An Introduction to New Media* (London: Berg, 2009).

33. Leopoldina Fortunati, James E. Katz and Raimonda Riccini, eds., *Mediating the Human Body: Technology, Communication, and Fashion* (Mahwah, NJ: Lawrence Erlbaum, 2003); Jane Vincent and Leopoldina Fortunati, eds., *Electric Emotion: The Mediation of Emotion via Information and Communications Technologies* (New York: Peter Lang, 2009); Ingrid Richardson, "Pocket Technospaces: The Bodily Incorporation of Mobile Media," *Continuum* 21, 2007: 205–215; Ingrid Richardson, "Ludic Mobilities: The Corporealities of Mobile Gaming," *Mobilities* 5, 2010: 431–447.

34. See, for instance, Ingrid Richardson and Rowan Wilken, "Haptic Vision, Footwork, Place-Making: A Peripatetic Phenomenology of the Mobile Phone Pedestrian," *Second Nature* 1, 2009: 22–41; Hjorth, *Mobile Media*.

35. Many commentators liken the mobile to the "Swiss army knife," including Hjorth, *Mobile Media* and Ling, *Mobile Connection*. On the dilemma of the Swiss army knife, see Donald A. Norman, *The Invisible Computer* (Cambridge, MA: MIT Press, 1998), discussed in relation to mobiles by Klaus Goldhammer, "Devices—Of Razors and Mobile Phones," in Jo Groebel, Eli M. Noam and Valerie Feldmann, eds., *Mobile Media: Content and Services for Mobile Communications* (Mahwah, NJ: Lawrence Erlbaum, 2006).

36. Bruno Latour, *Aramis, or the Love of Technology*, trans. Catherine Porter (Cambridge, MA: Harvard University Press, 1996).

37. Giovanni de Micheli, Rolf Ernst and Wayne Wolf, eds., *Readings in Hardware/Software Co-Design* (San Francisco and London: Morgan Kaufmann, 2002).

38. On the shaping of software, see David G. Messerschmitt and Clemens Szyperski, *Software Ecosystem: Understanding an Indispensable Technology and Industry* (Cambridge, MA: MIT Press, 2003); Matthew Fuller, ed., *Software Studies: A Lexicon* (Cambridge, MA: MIT Press, 2008).

39. Harmeet Sawhney, "Innovations at the edge: the impact of mobile technologies on the character of the internet," in Gerard Goggin and Larissa Hjorth, eds., *Mobile Technology: From Telecommunications to Media* (New York: Routledge, 2009), 105–117.

40. Atsushi Akera and Frederik Nebeker, eds., *From 0 to 1: An Authoritative History of Modern Computing* (Oxford and New York: Oxford University Press, 2002).

41. On computer cultures, see Timothy Leary, *Chaos & Cyberculture* (Berkeley, CA: Ronin, 1994); Susan Leigh Star, ed., *The Cultures of Computing* (Oxford UK and Cambridge, MA: Blackwell, 1995); Andrew Herman and Thomas Swiss, eds., *The World Wide Web and Contemporary Cultural Theory* (New York: Routledge, 2000); Jon Dovey and Helen W. Kennedy, *Game Cultures: Computer Games as New Media* (Maidenhead, England and New York: Open University Press, 2006); Christa Sommerer, Laurent Mignonneau and Dorothée King, eds., *Interface Cultures: Artistic Aspects of Interaction* (Bielefeld: Transcript; London: Transaction, 2008).

42. Ian H. Witten, *Communicating with Microcomputers: An Introduction to the Technology of Man-Computer Communication* (London and New York: Academic Press, 1980).

43. Steven G. Jones, ed., *Cybersociety 2.0: Revisiting Computer-Mediated Communication and Community* (Thousand Oaks, CA: Sage, 1998).

44. "Why do I need Viber?" http://www.viber.com/ (accessed 10 May 2011).

45. Gerard Goggin and Kate Crawford, "Moveable types: the emergence of mobile social media in Australia," *Media Asia Journal*, 37, 2010.

46. On the recurring dream of videotelephony, see Gerard Goggin, *Cell Phone Culture: Mobile Technology in Everyday Life* (London and New York: Routledge, 2006).

47. Apple, "FaceTime: Phone calls like you've never seen before," http://www. apple.com/iphone/features/facetime.html (accessed 10 May 2011).
48. Daniel Eran Dilger, "Apple's iPhone 4 FaceTime doesn't need a mobile signal to work," *Apple Insider*, 29 June 2010, http://www.appleinsider.com/ articles/10/06/29/iphone_4_facetime_doesnt_need_a_mobile_signal_to_ work.html
49. Julian Morel and Christian Licoppe, "Studying mobile video telephony," in Monika Büscher, John Urry and Katian Witchger, eds., *Mobile Methods* (New York: Routledge, 2010), 164.
50. Alasdair Allan and Pete Warden, "Got an iPhone or 3G iPad? Apple is recording your moves," *Radar*, 20 April 2011, http://radar.oreilly.com/2011/04/ apple-location-tracking.html
51. Francesco Lapenta, "Geomedia: on location-based media, the changing status of collective image production and the emergence of social navigation systems," *Visual Studies* 26, 2011: 14–24.
52. See, for example, Adriana de Souza e Silva and Daniel M. Sutko, "Theorizing locative technologies through philosophies of the virtual," *Communication Theory*, 21, 2011: 23–42; Adriana de Souza e Silva and Jordan Frith, "Locative mobile social networks: mapping communication and location in urban spaces," *Mobilities* 5, 2010: 485–505; and Ingrid Richardson and Rowan Wilken, "Parerga of the third screen: mobile media, place and presence," in Rowan Wilken and Gerard Goggin, eds., *Mobile Technology and Place* (New York: Routledge, 2012).
53. Ilpo Koskinen, *Mobile Multimedia in Action* (New Brunswick, NJ: Transaction, 2007).
54. For instance, see Louis Wirth, "Consensus and mass communication," *American Sociological Review*, 13, 1948: 1–15, and the commentary offered by Eric W. Rothenbuhler in his "Community and pluralism in Wirth's 'Consensus and Mass Communication'," in Elihu Katz, John Durham Peters, Tamar Liebes and Avril Orloff, eds., *Canonic Texts in Media Research: Are There Any? Should There Be? What About These?* (Cambridge: Polity Press, 2003), 107–120.
55. Marshall McLuhan, *Understanding Media: The Extensions of Man* (London: Routledge & Kegan Paul, 1964).
56. James E. Katz, *Magic in the Air: Mobile Communication and the Transformation of Social Life* (New Brunswick, NJ: Transaction, 2006).
57. On the politics of apps, see Gerard Goggin, "'Ubiquitous apps: politics of openness in global mobile cultures," *Digital Creativity*, 22.3, 2011: 148–159.

3   The iPhone Moment, the Apple Brand, and the Creative Consumer
From "Hackability and Usability" to Cultural Generativity

Jean Burgess

INTRODUCTION

In this chapter I position the iPhone as a moment in the history of cultural technologies. Drawing predominantly on advertising materials and public conversations about other "moments" in the history of personal computing and focusing on Apple's role in this history, I argue that the design philosophy, marketing and business models behind the iPhone (and now the iPad) have decisively reframed the values of usability that underpin software and interface design in the consumer technology industry, marking a distinctive shift in the history and contested futures of digital culture.

In the familiar showbiz style with which any new Apple product is launched, the iPhone was first unveiled by Steve Jobs in front of thousands of people in January 2007, at the annual Macworld Expo.[1] Months of ritualized anticipation, rumors and speculation about Apple's next move reached their climax as Jobs strode onto the Macworld stage clad in his characteristic jeans and turtleneck and proclaimed, "We're going to make some history here today." Following a series of lesser announcements (updates to the Mac, the launch of the AppleTV) came the iPhone's big reveal. "This is a day I've been looking forward to for two and a half years," said Jobs. "Every once in a while a revolutionary product comes along that *changes everything*." Right from the outset, then, the meaning of the iPhone was structured around the discourse of revolutionary transformation—for the mobile phone, for personal technology and for the Apple brand.

In his keynote, Jobs himself situated the iPhone in terms of a series of equally "revolutionary" moments in the history of the company. In 1984, he said, "Apple introduced the Macintosh and changed the computer industry"; then in 2001, "Apple introduced the iPod and changed the entire music industry." "Well, today," he said, "we're introducing three revolutionary products of this class." "The first one is a widescreen iPod with touch controls. The second is a revolutionary mobile phone. The third is a breakthrough internet communications device." The punch line, of course, is that "These are not three separate devices. This is one device. And we are calling it iPhone. Today Apple is going to *reinvent the phone*."

As Jobs quite deliberately pointed out, the launch of the iPhone saw Apple enter the mobile phone market, at the same time marking a significant moment of technological and media convergence between the phone and the personal computer. Indeed, the launch of the iPhone was the most emphatic assertion to date of the convergent future of the mobile phone, media consumption, social networking and personal computing—a convergence that in the years since the iPhone's launch has become normalized via a large and increasingly diverse smartphone market. There is no clearer demonstration of this shift in the logic of Apple as a corporation than the announcement, made at the same event, that the company was changing its name from Apple Computing to simply Apple, Inc.—the company was no longer only in the computer business; it was in the consumer technology business.[2] Somewhat disingenuously, given the computational power upon which all of Apple's devices rely, Jobs explained, "The Mac, iPod, AppleTV and iPhone. Only one of those is a computer. So we're changing the name." The iPhone is thus not just a highly popular mobile media device that happens to be produced by Apple: the iPhone "moment" also repositioned the Apple brand. As I discuss below, the iPhone moment also invited us, as users, to be repositioned in relation to the technologies we integrate into our everyday lives. This repositioning is only part of a longer history of struggles among various visions for, and market realizations of, the potential of personal computing; a history that, with the iPhone, the mobile phone now enters.

HACKABILITY VS USABILITY

Far from the *oohs* and *aahs* of the Macworld crowd, though, the iPhone moment was and continues to be a site of significant controversy around technological design and use, as is the Apple brand and its associated products more generally. Debates abound about the extent to which the iPhone is an "open" or "closed" device, and what the political or economic implications might be in either case, mirroring high-stakes debates about the future of the internet and personal computing more broadly. One of the first scholars to articulate these broader political concerns about the iPhone being a closed device was Jonathan Zittrain in his influential book *The Future of the Internet—And How to Stop It*.[3] Zittrain characterizes the iPhone as a mere "tethered appliance," in contrast to more generative, user-configurable and programmable "platforms" like the computer.[4] The broader issues for Zittrain concern what he sees as a trend toward corporate enclosure of media and internet technologies, gradually removing technological agency from the hands of the user and centralizing control in the hands of corporations like Apple. For Zittrain, as an appliance the iPhone allows users to undertake only prescribed actions, making technology less "generative" of new uses and possibilities, and therefore hampering both innovation and democracy. Even since the launch of the App Store in 2008—clearly a "generative" affordance of the iPhone and its operating system iOS—Zittrain's concerns have been mirrored in debates across a wide range of media, but especially among technology and software enthusiasts and some developers. See, for example, the many articles discussing the topic of the iPhone's "closed" nature as a platform (and the extensive—and heated—comment threads that inevitably follow) in leading technology news websites;[5] or the now-infamous articles that appeared upon the launch of the iPad where (usually male) technology enthusiasts attempted to make sense of its hybrid position (not quite computer; more than a phone) by arguing it was a device their "moms" might like.[6]

In microcosm these concerns represent ongoing struggles and complex articulations between the ethics and values of "hackability" (founded in but extending beyond self-nominated hacker movements) and "usability" (which is equally founded in democratic, but far more market- and consumer-driven, principles). These two sets of values can be understood as occupying opposite points on a continuum: *hackability* is a state under which a given technology is open-ended, manipulable and affords complex experimentation with an accompanying level of difficulty and expertise, and at the other end is extreme *usability*—where a technology affords easy access to a pre-determined set of simple operations, often via intuitive, "friendly" interfaces.

The apparent opposition of usability and hackability is linked to the unresolved tensions between the ideologies of critical modernism and postmodern populism. According to these respective ideologies, hackability, as an ideal, permits rational mastery and understanding of the technical reality of machines, but in reality, it is only the technical *avant-garde* (like computer scientists or hacker subcultures) who achieve this mastery. In contrast, usability permits popular access without the need for esoteric knowledge, by creating pleasurable surface interfaces that automate operations over the underlying layers of technology. Thus, usability opens up technologies to the masses (or, pejoratively, "dumbs down" technologies to the extent that even "your mom" can use them). The interplay and tensions between these ideologies (in short, radical democracy vs market-driven populism) are inseparable from the history of cultural technologies.

THE NORMALIZATION OF USABILITY AND THE DOMESTICATION OF PERSONAL COMPUTING

In contemporary developed societies, the everyday relationship of mainstream consumers to computer use has for some time been naturalized—it is quite literally mundane. But this ordinary or everyday positioning of the personal computer represents a dramatic shift from the "original" meaning of the technology—and it is a shift that took place over only three decades of the late twentieth century. The process underlying this shift is known as "domestication" and it involves not only manufacturers and marketers, but also the participation of consumers or users as they make sense of, adapt and integrate new technological objects and practices into their everyday lives, both inside and outside the home.[7] Roger Silverstone and Leslie Haddon argue that consumption and use are "equally essential components of the innovation process" but that the market (including advertising materials like those I discuss in this chapter) plays a crucial role as well. Their concept of "catching the consumer" encompasses the duality of the relationship between technology and the market: on the one hand, technology is central to consumer capitalism; on the other, the market is central in "defining the status and meaning of technology."[8] Indeed, "It is the market, in the form of commodification, which intrudes as a crucial component in the definition of technology precisely at the point where design and use confront each other."[9]

From the 1950s until the early 1970s computers were articulated to science and the military:[10] they were mostly institutionally owned and controlled by a small, closed group of expert scientists and technicians— far from the lifeworlds of ordinary people—but alongside the monolithic computer industry, a satellite subculture of hackers emerged. This early hacker movement, which was primarily responsible for the innovations that evolved into what was then called the "small computer," was characterized by an intense, hyper-rational engagement with technology, usually with the aim of "getting the system to do something other than the function intended for it by its designer."[11] The movement was populated by highly individualistic, almost always male mavericks connected to research and science institutions such as MIT who competed and, to some extent, collaborated to find the most "elegant" solutions to computational problems.[12] The hackers, like early photographers, were both experts and enthusiasts—operating in a highly individualistic way but philosophically engaged with the social or liberatory potential of computing. The public good for personal computing as imagined by hacker culture tended to be ubiquitous access and the democratization of education and information; less emphasis was placed on personal expression or entertainment, even though for garage enthusiasts, hacking clearly had its pleasures—the apocryphal tales of young Apple founders-to-be Steve Jobs and Steve Wozniak's early hijinks and later development of the Apple I in a suburban garage, as members of the Homebrew Computer Club, are paradigmatic of this ethos.

The following quote from the *Homebrew Computer Club Newsletter* published in 1975—the first year of the club's operation—neatly encapsulates the more populist versions of the hacker ethic:

By sharing our experience and exchanging tips we advance the state of the art and *make low cost home computing possible for more folks* . . . Computers are *not magic*. And it is important for the general public to begin to *understand the limits* of these machines and that humans are responsible for the programming [my emphasis].[13]

The idea of user agency embedded in the hacker ethic is still alive and well today in the philosophy of the open source software movement, as well as being reconfigured in the more playful "DIY" technology and craft movement metonymically represented by the motto of *Make Magazine*: "If you can't open it, you don't own it."[14]

The beginning of the personal computer "revolution" in the late 1970s and early 1980s, just as Apple was really getting going, was a period of technological and discursive diversity and instability. The market was crowded with competing machines, each with its own operating system and user interface. The early 1970s saw ideological contests erupt over the meanings and purposes of the small computer: on one hand, they were viewed as delivering power to the people (the libertarian or radical hacker ethos); on the other, discourses of efficiency and rationalization emanated from the business world. The early 1980s marked the beginning of the convergence of play and fun with literacy, technological mastery and technical knowledge, as games, music and graphical applications began to feature more prominently. Early personal computing ideology was an unstable mix of hacker ideology and play—the emphasis for the ideal user addressed by advertisements for personal computers was on writing one's own programs, games or music, rather than using provided applications.

This is particularly marked in the case of early personal computers like the Amiga, Atari and Commodore 64, whose marketing focused on the assumption that users would want to program music or games rather than just "passively" play or use those created by others, and that the knowledge of computers *as computers* (and not as interfaces) was important and valuable. The Amiga 1000, for example, was a multimedia personal computer with exceptional graphics performance for the time. In a television advertisement for the machine, while images of graphs, spreadsheets, games and music composition appear on the screen, the voiceover frames the product as a way to "stand out from the crowd" in a competitive world. The Amiga, the audience was told, "gives you undreamed of creative power" to "work faster and more productive" [sic] and was "the first personal computer to give you a creative edge."[15] In this advertisement and those like it from the period, individual business success and productivity are articulated to personal creativity and exceptional "ability" in a style resonant of the "success ethic" that Elizabeth Traube[16] identified in 1980s Hollywood films like *Ferris Bueller's Day Off*, *All the Right Moves* and *The Secret of My Success*, arguing that the "fantasy embedded in the commercially successful success stories" of these films contributed to "the making of the new middle classes."[17]

While the success imperative as a reason for computer take-up was directed at young people, especially boys, in a parallel and slightly later 1980s trend, ordinary domestic users—particularly women—were schooled to adopt the computer as a benign and useful home appliance. Lori Reed traces the processes through which computer technologies were disarticulated from science (or science fiction) and re-articulated to the home, thereby discursively transforming them from "cold, distant military war machines" into "friendly" home appliances,[18] producing "eventually accepted linkages to the home, family, business, and pleasure, such that today computer technologies are 'naturally' integrated into many people's daily lives."[19] The mechanisms of this reconfiguration in the domestication of the personal computer included advice columns, marketing materials and public dialogue about computers and the range of their risks, benefits and potential applications. Part of the move to mass-market and domesticate the personal computer, as with photography, involved feminization—the computer was discursively "softened." The cultural and industrial emphasis on women's attitudes to computers was key for the "cajoling" of computers into the hands of women and into the home. At the beginning of the domestication process, the reluctance to take up computing was constructed as a "phobia" by the media.[20] Once the process of widespread domestication was complete and home users had begun to integrate computing into their everyday lives in ways that were not prescribed, there was a shift into discourses of computer "addiction"—once again centered around women—and we can see a similar pattern in discourses of addiction and risk around social and mobile media use (particularly for young people) today.[21]

One of the most significant developments during this time was the normalization of the graphical user interface (GUI) in the mid-1980s. This triumph of the interface designers (the postmodern populists) over the hardware-and-code-oriented hackers (the modernist avant-garde) resulted in a complete redefinition of technological transparency. Whereas for the hackers, transparency meant visibility and openness at all levels of hardware and code so that users might learn and fully master the computer, "user-friendly" interface design principles redefined transparency to mean the invisibility of *all* technological layers, leaving only the GUI so that there was nothing standing between the will of the user and the task for which he or she wished to use the computer. This redefinition of "transparency" is now normative, at least in interface design, and reflected in Jay Bolter and Diane Gromala's statement that:

> When designers set out to define an interface for an application . . . they usually assume that the interface should serve as a transparent window, presenting the user with an information workspace without interference or distortion. They expect the user to focus on the task,

not the interface itself . . . If the application calls attention to itself or intrudes into the user's conscious consideration, this is usually considered a design flaw.[22]

The iPhone's touchscreen, gestural interface and icon-based operating system couldn't be more closely aligned with this notion of radical transparency, and the underlying architecture of Apple's iOS couldn't be more carefully hidden from us as users. Yet it is this philosophy of transparency, crystallized in the ascendancy of the GUI in the 1980s, that has led to technological access for non-specialist users. While the ethic of seamless usability represented by the GUI limits, shapes and constrains popular access to technology, it is also its very condition of possibility. This more positive view of usability—as a means of making complexity accessible without reducing it to simplicity—is expressed in Donald Norman's germinal work *The Invisible Computer*,[23] which has been extremely influential in the field of human-computer interaction (HCI).

The domestication of personal computing represents a major cultural and technological transition that was achieved not only by designers and technologists, but also by markets, advertising and media discourse. The meaning and the uses of the computer were transformed: from elitist but hackable war machine to usable home appliance. Today, this latter trajectory has culminated in mundane, mobile and ubiquitous personal technology, whose profound complexity is harnessed for everyday use via software interfaces that mediate apparent transparently between the will of the user and the functionality of the device.

THE EVOLUTION OF THE APPLE BRAND

The 18-page brochure insert that introduced the GUI and mouse-equipped Macintosh computer in 1983 told us that "of the 235 million people in America, only a fraction can use a computer."[24] Page two announced, "Introducing Macintosh. For the rest of us." The famous "1984" commercial that announced the impending release of the Macintosh emphasized the freedom from corporate conformity (represented by IBM) that would be delivered by the Macintosh's superior usability. The commercial aired on January 22, 1984, during a break in the third quarter of the Super Bowl. It portrayed a young heroine wearing orange shorts, red running shoes, and a white tank top with a picture of Apple's Macintosh computer on it, running through a dark world populated by drones, and eventually hurling a sledgehammer at a television image of Big Brother (a thinly veiled reference to IBM). The advertisement concluded with the message: "On January 24th, Apple Computer will introduce Macintosh. And you'll see why 1984 won't be like '1984'." From the beginning, Apple focused on this very particular and deliberate construction of human-centered and populist usability above all else.

From the late 1980s, Apple capitalized on the perception of the superior performance and graphics processing power of the Mac (in part facilitated by Apple's partnership with Adobe software), concentrating on the creative professionals market (such as graphic designers and musicians), and by association also extending an invitation to ordinary users who aspired to creative practice. This focus effectively drew on the dominant cultural ideology of the period—the success narrative. Indeed, during this period, "the power to be your best" was the usual slogan in Apple's television advertising.

Apple's late 1990s "Think Different" advertising campaign, which coincided with the return to Apple of co-founder Steve Jobs and the release of the PowerMac, reinforced the construction of both the Apple brand and Apple's users as exceptionally creative and non-conformist. The one-minute television commercial that was at the center of this campaign featured black-and-white video footage of creative, iconoclastic and well-known historical figures like Albert Einstein, Pablo Picasso and Muhammad Ali.[25] The dramatic voiceover began as follows:

Here's to the crazy ones.
The misfits.
The rebels.

It went on to highlight—in a measured, poetic cadence—the qualities of individualism, vision and extraordinary uniqueness associated with creative and innovative people before concluding:

We make tools for these kinds of people.
While some may see them as the crazy ones, we see genius.
Because the people who are crazy enough to think that they can
    change the world are the ones who do.[26]

The advertisements that make up the "Think Different" campaign all relied upon a process of metaphoric transference: like Einstein, Picasso and Ali, Apple is creative, non-conformist and innovative; through the slogan "think different" these qualities of distinction are offered to the user. A distinctly Romantic construction of creativity permeates these advertisements: those who "think different" are crazy, nonconformists and geniuses—there is nothing ordinary about them; however, it is possible for ordinary people, with the aid of the right tools, to aspire to great creative achievements.

In a move that explicitly differentiated the Apple Macintosh from the generic "beige" Windows PC, Apple targeted the mass market with renewed vigor with the release of the iMac in 1998. The design of the machine, and its representation in advertising, marketing and media discourse, represented the radical aestheticization of ordinary personal computing—the iMac, with its transparent all-in-one case in a range of bright colors (with names like tangerine and grape appealing to the sensation of taste), housing both monitor and computer, was explicitly designed to create the impression that a convergence of fun, pleasure and processing power had occurred and that it was unique to the iMac. It was more than a "pretty" material object, however; the iMac in combination with the Mac OS also represented a radical ethic of usability characterized by the seamless integration of hardware, operating system and applications. The Apple slogan that captures this repositioning of the brand proclaimed that computing, iMac style, was "chic, not geek."[27]

The early 2000s advertising campaigns that coincided with Apple's mass-marketing of the iMac for a general rather than specialized user community advocated a more active, even subversive form of consumption; however, there was little sense in which the user was constructed as productively creative. In the "Rip. Mix. Burn" television advertisement that aired in the US to promote the release of iTunes, a young male user seated in an empty auditorium demonstrates the ease of creating music playlists by instructing the real-life celebrity musicians assembled on stage as to which songs they should perform and when. This advertisement explicitly reconfigures the practice of music consumption as the active exercise of knowledgeable taste and consumer agency. With the extensive consumer adoption of the iPod and the success of the "Switch" campaign in the early to mid-2000s, where celebrities and "ordinary people" alike related their personal accounts of the many advantages of owning a Mac, the Apple brand community became even less exclusive. The ideal Apple user was becoming increasingly "ordinary" while continuing to represent the ideal creative consumer from the perspective of the consumer technology industries.

Prefiguring the positioning of the iPhone (and iPad) as "platforms," much of the advertising for the various incarnations of Apple's laptop computers and iPods situated the devices as relatively neutral tools that would allow for a proliferation of applications and taste choices customized to the identities of a carefully diverse universe of consumers—marking another significant break away from a "hacker" ethic of use and towards a consumer-oriented ethic of endless, seamless usability. Continuing in this direction, the 2006–2007 series of twenty "Get a Mac" television advertisements[28] directly personified the Macintosh and the IBM-compatible PC. The advertisements featured actor Justin Long as a Mac and author, and humorist John Hodgman (The Daily Show) as a PC (presumably running Windows). In each advertisement, Long introduced himself as a Mac and Hodgman introduced himself as a PC, then a particular aspect of computing was set up as the basis for comparison.

In these advertisements, the result of the personification of the brands was that the boundary between the lifeworld of the user and the technology is dissolved; complete convergence is achieved. Both of the personified computers were white, male and vaguely middle-class North Americans. Neither was marked as culturally "different"; the Mac was simply a better class of geek—breezily confident, even smug, and seamlessly integrating everyday life, leisure, creativity and work; the PC was arrogant, spiteful and jealous, trapped in last century's models of productivity, and appeared physically bloated, with bad hair and an unfashionable business suit.

In the 2006–2007 "Get a Mac" campaign, which targeted potential "switchers" from the PC to the Apple computer market, one of the key differentiating factors was the iLife suite of creative software applications. Apple's iLife software suite came bundled with all new Apple computers and was a central part of Apple's marketing strategy for its personal computers.[29] This software suite and the marketing around it were also representative of a relatively recent shift in Apple's branding strategy that constructed its users as effortlessly creative rather than extraordinarily so. In the list of fourteen reasons that the target audience for the campaign "will love a Mac," the promise of access to creativity offered by the machine—"you can make amazing stuff"—ranked second only to the promise of effortless usability—"it just works."[30] Illustrating this new convergence of ordinariness, creativity and the particular construction of technology represented by the Apple brand, the iLife promotional videos that accompanied the launch of the software suite in 2004 explicitly constructed creativity as something both everyday and cool; the "ideal" user was directly represented, whereas the earlier campaigns did not directly locate a fully formed capacity for creativity in the ordinary user. The marketing information that accompanied the launch of iLife in January 2003 highlighted the link between usability and creative expression:

> Let your imagination soar: The iLife software applications let you do fun, creative things with your pictures, music and movies in ways that PC users can only dream about—and then you share your joy with family and friends every which way, from email and the Internet to print and DVD.

> And you can do all these things and more quite nicely without thumbing through a manual. It's all part of the iLife experience.[31]

The constructions of both creativity and usability at work here rely on a particular notion of technological transparency: the software lets you "do fun, creative things" that are a direct expression of an inner urge to create; the need to develop specialist expertise in order to do those creative things is erased, because they can be achieved "without thumbing through a manual".

The iLife suite represented a key moment in the development of the creative consumer as an ideal user of technologies designed for everyday use. It also represented a key point of convergence between play and productivity in the ethos of everyday technology use, and a blurring of the divide between professional creative production and everyday technological or media consumption. The discursive construction of vernacular creativity in relation to the Apple brand therefore contributes to the production of a new ideal personal computer user who does not need to "master his tools," as Ivan Illich might say,[32] but rather enjoys using them playfully, producing and repurposing content that emerges seamlessly out of the articulation of fun, creativity, technology and his or her everyday experience.

The emphasis on seamless creative production enabled by easy-to-use and attractive technologies did not go uncontested. Of all the applications included in the iLife suite, it was the music composition software GarageBand that attracted the most discussion and debate. The themes of the discussion followed familiar patterns—high levels of enthusiasm bordering on hyperbole from the Mac user community and *Wired* magazine, in turn opposed by the expert discourse of music software aficionados.[33] The "experts" were often vitriolic in their arguments that GarageBand, because it was so easy to use and relied heavily on dragging and dropping loops, rather than requiring the user to compose, or even sample, her own music, might cause a "flood" of banal and poorly produced music. Even the most loyal "amateur" users soon found the limits of the application as they attempted to experiment with musical composition and production beyond "dragging and dropping" the supplied loops, but individual users and user communities began searching for and developing workarounds, many of which were later superseded by Apple's integration of the features users had felt were lacking.

Several online GarageBand user communities also developed—places where musicians could share their finished tracks, as well as critique, teach, learn and collaborate with other users.[34] These online communities provided free space to upload and share GarageBand compositions; commenting and rating systems, forums and the integration of Creative Commons (or Creative Commons style) licenses to afford legal remixing and collaboration among users. The practices of these user-initiated GarageBand communities exceeded the idea of the Mac user as an individual "creative consumer," encompassing collaboration, critique, reworking and remixing one's own as well as others' work. Even the most simplified usability doesn't necessarily mean "dumbing down": it can open up opportunities for greater participation, including critical participation, learning and the development of cultural, not only technical, mastery.

THE iPHONE MOMENT AND BEYOND

In the decades-long history of personal computing and Apple's role within it, the iPhone moment was a particularly marked break. The iPhone saw not only a reconfiguration of the mobile phone, but also of the Apple brand, and in particular how Apple's design practices, advertising and the materiality of software position us as users and as cultural agents. At the "iLife moment," bona fide creativity became naturalized as an attribute of the ideal ordinary user of the Apple personal computer, while technological mastery was removed as a requirement. However, the iPhone moment marks a different kind of shift in the way the user is positioned—indeed, in most advertisements for the iPhone (and now the iPad), the brand attributes of the device are no longer transferred to the user as they were in earlier advertising campaigns.

Instead, the marketing of the iPhone consistently represents the device as a kind of magical object with which we are physically intimate, and which responds to our interior thoughts and desires with the mere touch of a finger, but the creativity of the user as a unique individual is no longer really foregrounded.[35] We couldn't be further from "Think Different"—Apple no longer emphasizes market differentiation, instead asserting mass market ubiquity—an iPhone in every hand (and pocket). We also couldn't be further from the ideals of the Homebrew Computer club—Apple technologies *are* in fact represented as magic; we *don't* need to know how they work or to understand them "as computers":[36] to quote an earlier Apple slogan, the iPhone and iPad are magic, usable, and endlessly configurable, but we don't need to worry about how, because they "just work."

And in a further difference from the "power to the people" ethos of the earlier "Think Different" moment, in the advertising and marketing surrounding the iPhone moment, the devices themselves are not so much *personalized* as *anthropomorphized*—giving new meaning to Jobs' long-standing habit, which flows through to all official Apple discourse, that Apple products be referred to without a definite article—not "a" Macintosh, but "Macintosh"; not "the iPhone", but "iPhone." iPhone (and iPad) users are not represented as full unique humans, but as mere gestural interfaces (fingers and hands) and software preferences; in the most recent iPad television advertisements, for example, individual relationships with the device are used only to emphasize its universal adaptiveness signified by the endless possibilities of its usability. One recent TV commercial with the tagline "iPad isn't just one thing" cycles through a range of markedly everyday uses associated with different generic social identities, represented by apps—it can be used (by Mom) for recipes; by the kids for playful learning; or for drawing, games, reading and social networking.[37]

This strangely mundane adaptiveness with which the iPhone and iPad are now associated is indicative of a recent and wide-ranging shift (in relation to mobile media, traceable back to the launch of the App Store in 2008) to the paradigm of the *platform*—a shift that is mirrored more broadly in the ways "Web 2.0" platforms like YouTube are positioning themselves and their users. This broader paradigm shift is discussed in depth by Tarleton Gillespie, who has traced the multiple and often contradictory ways that online content and service providers are framed as "platforms," and the political consequences for the way users are positioned in relation to these companies' business models.[38] In more recent advertising campaigns for the iPhone, Apple has strongly emphasized the device as a platform for the possibilities generated by the millions of applications (apps) available to users—covering everything from gaming to workplace productivity to creative photography, social networking, and navigation. From the "There's an App for That" campaign to the more recent "If you don't have an iPhone, you don't have the App Store," in each advertisement the endless *generativity* of the iPhone as platform is positioned front and center.

Elsewhere, I have argued that from the very beginning, YouTube's architecture and possible uses were underdesigned and therefore "underdetermined," producing conflicts and tensions around its politics and its future, but also allowing for an extremely wide range of possible uses and future developments.[39] This relative openness, despite YouTube's corporate ownership (and increasing regulatory control), makes it a *culturally generative* platform. The iPhone (by which I mean the device itself, its operating system and the ecosystem of software development that is channeled through the App Store) is by contrast somewhat overdetermined; it is far more tightly and consistently managed than YouTube was, at least in the first few years of its evolution.[40] The iPhone, as an appliance, is a "closed", tightly controlled device, and (with the exception of its susceptibility to "jailbreaking"), it is not hackable either as hardware or as software. But it is also highly "usable", opening up possibilities for mobile media consumption, play and self-expression to an immensely broader population than the personal computer ever did. And as a *platform*, it provides an environment within which content, games and software innovators can introduce an extremely wide range of further possibilities for the iPhone's use, despite—or even thanks to—the tightly controlled ecosystem of the App Store (discussed in detail by John Banks in this volume). What the iPhone lacks in technological "hackability," therefore, it makes up for in social and cultural "generativity," thanks to its usability and the proliferation of apps that extend its functionality, and more importantly thanks to the creative, social and communicative activities of its millions of users who have integrated the iPhone, as a platform, into their everyday lives.

ACKNOWLEDGMENTS

I am extremely grateful to John Banks, Larissa Hjorth and Ingrid Richardson for their helpful comments on earlier drafts of this chapter.

NOTES

1. Jobs' keynote is available in video format via iTunes at http://itunes.apple.com/us/podcast/apple-keynotes/id275834665, and several copies are available on YouTube. It was also transcribed live by writers for various technology publications: see for example Ryan Block, "Live from Macworld 2007:

Steve Jobs keynote," *Engadget,* http://www.engadget.com/2007/01/09/live-from-macworld-2007-steve-jobs-keynote/ (accessed 22 April 2011).

2. Matthew Honan, "Apple Drops 'Computer' from Name," http://www.mac-world.com/article/54770/2007/01/applename.html (accessed 21 May, 2011).
3. Jonathan Zittrain, *The Future of the Internet—And How to Stop It* (New Haven and London: Yale University Press, 2008).
4. Ibid., 2–3.
5. Mike Masnick, "From Closed to Open: iPhone App Developer Skepticism Highlights Platform Trajectory," *Techdirt,* 19 July 2009, http://www.tech-dirt.com/articles/20090719/1514125593.shtml (accessed 14 April 2011).
6. Daniel Indiviglio, "Five Reasons Why Your Mom Wants an iPad," *The Atlantic Monthly,* 5 June 2010, http://www.theatlantic.com/technology/archive/2010/06/5-reasons-why-your-mom-wants-an-ipad/58185/ (accessed 15 April 2011).
7. Roger Silverstone and Leslie Haddon, "Design and the Domestication of Information and Communication Technologies: Technical Change and Everyday Life," In *Communication by Design: The Politics of Information and Communication Technologies,* in Robin Mansell and Roger Silverstone, eds. (Oxford: Oxford University Press, 1996), 44–74.
8. Ibid., 45.
9. Ibid., 59.
10. Graeme Kirkpatrick, *Critical Technology: A Social Theory of Personal Computing* (Aldershot: Ashgate, 2004), 25.
11. Ibid., 26.
12. Steven Levy, *Hackers: Heroes of the Computer Revolution* (New York: Dell, 1984).
13. "The Homebrew Computer Club," *Computer History Museum,* http://www.computerhistory.org/revolution/personal-computers/17/312 (accessed 14 April 2011).
14. "Owner's manifesto," *Make Magazine,* http://makezine.com/04/ownyourown/ (accessed 21 June 2011).
15. Commodore International, "Amiga 1000" (1985), currently archived at http://www.youtube.com/watch?v=F1R9tgDuuTo
16. Elizabeth G. Traube, "Secrets of Success in Postmodern Society," *Cultural Anthropology* 3(4), 1989: 273–300.
17. Ibid., 273.
18. Lori Reed, "Domesticating the Personal Computer: The Mainstreaming of a New Technology and the Cultural Management of a Widespread Tech-nophobia, 1964–," *Critical Studies in Media Communication* 2(17), 2000: 159–185.
19. Ibid., 163.
20. Ibid.
21. The discourse of risk as a way of organizing the meaning of children's inter-net use is so pervasive that it prompted a major multi-national survey project led by Sonia Livingstone and Leslie Haddon. The project, *EU Kids Online,* "centres on a cross-national survey of European children's experiences of the internet, focusing on uses, activities, risks and safety. It also maps parents' experiences, practices and concerns regarding their children's online risk and safety." See: http://www2.lse.ac.uk/media@lse/research/EUKidsOnline/Home.aspx.
22. Jay Bolter and Diane Gromala, "Transparency and Reflectivity: Digital Art and the Aesthetics of Interface Design," in Paul A. Fishwick, ed., *Aesthetic Computing* (Boston: MIT Press, 2006), 365.

23. Donald A. Norman, *The Invisible Computer* (Boston: MIT Press, 1999); see also his earlier work Donald A. Norman and Steven W. Draper, *User-Centred System Design: New Perspectives on Human-Computer Interaction* (London: Lawrence Earlbaum Associates, 1986).

24. Apple Computer, "Introducing Macintosh. For the Rest of Us," 2004, http://www.digibarn.com/collections/ads/apple-mac/index.htm (accessed 22 April 2011).

25. Including Albert Einstein, Bob Dylan, Martin Luther King, Jr., Richard Branson, John Lennon, R. Buckminster Fuller, Thomas Edison, Muhammad Ali, Ted Turner, Maria Callas, Mahatma Gandhi, Amelia Earhart, Alfred Hitchcock, Martha Graham, Jim Henson (with Kermit the Frog), Frank Lloyd Wright and Pablo Picasso.

26. Apple Computer, "Think Different," 1997, http://www.youtube.com/watch?v=No1MxAnHuJM (accessed 22 April 2011).

27. "iMaculate," *The Economist*, 7 January 1999, http://www.economist.com/node/181693?story_id=181693 (accessed 22 May 2011).

28. While Apple, Inc., has removed its own archive, many of the advertisements in the "Get a Mac" campaign have been curated by YouTube users. For example see http://www.youtube.com/watch?v=siSHJfPWxs8 (accessed 22 May 2011).

29. The suite contains software for imaging (iPhoto), video production and editing (iMovie), DVD creation (iDVD), and as of 2004 and 2005, respectively, music production (GarageBand) and web publishing (iWeb).

30. Apple Computer, "Apple—iLife," http://web.archive.org/web/20030108180049/www.apple.com/ilife/ (accessed 22 May 2011).

31. Ibid.

32. Ivan Illich, *Tools for Conviviality* (London: Marion Boyars, 1973).

33. Leander Kahney, "GarageBand Kicks Out the Jams," *Wired News*, 2004, http://www.wired.com/news/mac/0,2125,62204,00.html?tw=wn_culthead_4 (accessed 22 May 2011).

34. The most well-known of these were MacJams.com and iCompositions.com.

35. See, for example, the first four (US) TV commercials for the iPhone, archived by a YouTube user at http://www.youtube.com/watch?v=6lZMr-ZfoE4

36. "Homebrew Computer Club."

37. See http://www.youtube.com/watch?v=btfbIVGES1I

38. Tarleton Gillespie, "The Politics of 'Platforms," *New Media & Society* 12(3), 2010: 347–364.

39. Jean Burgess and Joshua Green, *YouTube: Online Video and Participatory Culture* (Cambridge: Polity Press, 2009).

40. For an extended and pragmatic discussion of the controversy surrounding the iPhone SDK and its status as a "closed" development platform, see James Grimmelmann and Paul Ohm, "Dr. Generative Or: how I learned to stop worrying and love the iPhone," *Maryland Law Review* 69, 2010: 910–953.

4 Ambient Intimacy

A Case Study of the iPhone, Presence, and Location-based Social Media in Shanghai, China

Larissa Hjorth, Rowan Wilken, and Kay Gu

INTRODUCTION

Each individual posting is essentially inconsequential but the overall effect is the gift of virtual co-presence in real time—ambient intimacy.[1]

Presence was a key concept in much of the early scholarship on the socio-cultural uses of the internet. Drawing on this literature and the work of contemporary theorists of presence, and building on existing scholarship that has explored presence in relation to mobile phones, in this chapter we explore how the introduction of devices such as the iPhone serves to further underscore the continued importance of presence as a crucial concept in an era of smartphones. The ongoing significance of presence in terms of smartphones is amplified with geomedia, whereby the social is overlaid with the geographic and electronic. According to anthropologist Daniel Miller, social media such as Facebook "suggest further movement towards the smartphone, rather than computer-based integration and convergence."[2]

This chapter takes up these issues by examining presence in the age of the iPhone, perhaps the most iconic of the current breed of smartphones, in relation to the Chinese location-based social (LBS) networking service, *Jie Pang*. In the LBS mobile game, *Jie Pang*, users can "check in" to online spaces they visit and win prizes. However, *Jie Pang* also allows users to notify friends about their location, and it is this secondary social motivation that we will see play out in this chapter. In order to explore *Jie Pang*, the chapter traverses four related areas. First, we contextualize the iPhone phenomenon by situating it within China's broader technoculture. Second, we provide some background to the main demographic of mainland China iPhone users, the *ba ling hou*. Born in the 1980s, the *ba ling hou* are the first generation to grow up in China's twenty-first century mediascape. This sets the context for the third part of the chapter: a case study of iPhone users, with a focus upon their deployment of the LBS, *Jie Pang*. The fourth and final part of the chapter reflects upon the types of co-presence that are afforded by LBS or "geomedia" services (as we prefer to call it here) like *Jie Pang*.

CHINA'S TECHNOCULTURES

Class (as a socio-economic category) continues to be one of the dominant factors influencing the types of emergent technocultures that are evident in contemporary China.[3] By technoculture we are referring to the ways in which culture is saturated by, and through, technologies. Class shapes, and is shaped by, the differing technocultures that are characteristic of China's rapid economic growth and dramatically shifting cultural landscape of the twenty-first century.[4] Through these technocultures, we can see numerous forms of mobility—socio-economic, geographic, generational, technological and psychological, to name a few—that complicate notions of age, class, place, and identity. This is particularly the case in relation to mobile media, whereby phenomenon such as *shanzhai* ("copy culture," especially apparent with the cloning of mobile media devices like smartphones) can be found in abundance. Once upon a time, the *shanzhai*'s loud ringtone signaled mobile media's poorer cousin; however, more recently, the divide between copy and original has been further complicated with the rise of smartphones.

In July 2011, three examples of *shanzhai* Apple Stores were found by an American blogger in China—within 72 hours of the post she had received over one million hits and global news media companies avidly picked up the story.[5] The story seemed to spearhead a tension between two very different technocultures—the *shanzhai* culture of China, on the one hand, and the Apple culture of the US, on the other. However, this opposition between the *shanzhai* and the original, China and the US, respectively, is much more complex, highlighting the localized nature of media practice. In China the *shanzhai* does not dilute the "aura" of such lifestyle objects as the iPhone; rather, like the Louvre's postcards of *Mona Lisa*, the copies play into the significance of the "original." The iPhone represents a particular lifestyle niche within Chinese technocultures—an issue that will be explored in more detail later in the present chapter.

China's role globally in the production and consumption of ICTs (information and communication technologies) has been most evident within the twenty-first century through the deployment of mobile phones[6] and the internet as sites for contestation and technonationalism.[7] In the case of mobile phones, they have had both material and immaterial dimensions and implications within the changing formations of Chinese technocultures. The mobile phone in general has provided much fuel as a symbol for, and of, the growing migrant working class.[8] As Jack Qiu's studies vividly identify, behind the role mobile phones play as commodities is another side— mobile phone production and its reflection of China's new working class.

In the case of the iPhone production, Qiu discusses the barbaric working conditions of Foxconn, home for much of Apple's production (see Qui's chapter in the collection). While the mobile phone has been dominated by discussions about the migrant working class, much of the literature on China's internet has focused on its role as a contested site between public opinion and government policies. As is the case in South Korea, the internet in China has provided a way in which to conceptualize struggles of democracy with social media such as blogs attracting much critical attention,[9] while seemingly "less political" media, such as social networking sites (SNSs) like QQ and Renren (China's equivalent of Facebook), have been relatively overlooked.[10] This is despite social networking services often being deemed threatening and thus banned by the government (in 2009, for instance, access to Facebook and Twitter was barred). However, with the rise of smartphones no longer just available to the middle class (especially thanks to *shanzhai*), and with the convergence between social media and mobile media having a long tradition (thanks to China's oldest social media, QQ, being mobile phone focused), we are seeing new ways in which the online is interweaving with the offline.

In 2009, for the first time, internet penetration rates in China surpassed the global average level with over 298 million users.[11] Over the last two years, China Internet Network Information Center (CNNIC) statistics have noted a sharp increase in lower-income and less-educated people becoming netizens. Just as access to the internet, via mobile phone or personal computer, differs within the divergent socio-economic groups, motivations for using the internet, in relation to types of activities and access routes (via mobile phone or personal computer), also differ. For many outside of big cities, the only access route onto the internet is via the mobile phone. This is reflected in some of China's social networking systems, such as the oldest, QQ, in which the numerous different services cater to diverse platforms and modes of access—particularly affording a specific version of mobile internet.

While technological infrastructure and access to the internet are still very much an urban preoccupation, significant changes have occurred in rural areas, with the number of rural internet users reaching nearly 85 million.[12] In short, the growth rate of rural internet usage now far exceeds that of urban internet usage. While the internet is mainly accessed via personal computer in urban city areas by the middle and upper class, the rest of China's predominantly working class demography deploys General Packet Radio Service (GPRS) for internet access via the mobile phone.

With this rise of internet access accompanying economic and geographic movements, an increased uptake of social media is also notable. Within China's net-scape, numerous types of SNSs—representing different classes, communities, and lifestyle clusters—can be observed. These SNSs predominantly fall into two technocultural types—sites such as Xiaonei/ Renren (akin to Facebook), Kaixin (used by female "white collars"), and

Microsoft Network (MSN) are often simultaneously open on someone's desktop, whereas Fetion and the aforementioned QQ can be accessed both via mobile phones and computers.

Within the broader Chinese mediascape, it is the *ba ling hou* generation who has dominated the usage of much of social media and geomedia—apart, that is, from QQ, which is deployed by both young and old in rural and urban settings. We will now provide some contextual background to the *ba ling hou* and their specific media practices.

THE MULTIPLE MOBILITIES OF *BA LING HOU*

In the complex and varied mediascape of China, Shanghai represents a particular version of Chinese media.[13] Over the last ten years, cities such as Shanghai have seen not only the implementation of new technologies in educational and work settings, but also an increasing trend of studying away from home. Born in the 1980s, the *ba ling hou* generation is the first to grow up with the internet and new and mobile media as part of everyday life, and the first to appropriate a form of mediated mobility. They are also a generation that has been allowed to move from the regions to the city to study—something unheard of for previous generations. Often from one-child families, these *ba ling hou* are incredibly close to their parents, and although geographic mobility for education is a given for this generation, there is a need for continual contact with friends and family back home. In studying away from home, the deployment of mobile media—especially for SNSs like QQ and Renren—has been integral to their negotiations of *home and away*.[14]

Many of the *ba ling hou* often board away from home at a university in Shanghai. Accustomed to a very close relationship with their families, the *ba ling hou* find that mobile media are helping not only alleviate loneliness and isolation caused by separation from their home and their parents, they also provide new ways in which to communicate with their parents. In addition, for many of the *ba ling hou*, an important part of new media knowledge acquisition, and their negotiation of home and away, is the transfer of this knowledge to their parents and even grandparents.

As the first generation to grow up in China's emerging net culture, the *ba ling hou* are a product of the first large-scale IT education project, initiated in 1994. This initiative involved the construction of a national network called CERNET (Chinese Education Research Network), followed by the rollout of EISS policies (an acronym for "Electronic Information Service System," or, in Chinese, "*xiaoxiaotong*"). Through these policies, the government orchestrated, over a ten-year period (2000–2010), the deployment of computers, networking equipment, and training programs to enable 90% of independent middle and primary schools to have access to the internet, accompanied by the provision of online content to be shared amongst teachers and students.[15]

The first to benefit from the EISS, the *ba ling hou* generation is now positioned in a particular narrative of progress in which technologies have played a central role. They are a generation that has high media literacy and, like their Western counterparts, view the internet as essential infra-structure for everyday life. Through this IT literacy, and against the wider backdrop of China's growing economic prosperity and growing middle class, many of the *ba ling hou* generation can travel—often for study. This link between the government's new media education initiatives, the provision of better access to new media technologies, and students' high levels of new media literacy, have allowed the *ba ling hou* to negotiate a sense of mobility unimaginable for previous generations. Thus, as the product of technology and education reforms, as well as China's newly found socio-economic and ideological prowess, the *ba ling hou* enjoy the benefits of online connectivity for study and social activities, including access to such fashionable consumer items as the Apple iPhone.

THE iPHONE IN CHINA

While locations such as Tokyo, Japan, have been accustomed to internet via mobile media for over a decade, with popular devices such as NTT DoCo-Mo's i-mode, many other locations, such as Chinese cities like Shanghai, have experienced comparatively slow take-up. However, in the last couple of years, the rise in popularity in China of smartphones—devices that respond to a desire for "seamless" mobile internet and telecommunications convergence—has continued unabated. Key to this broader fascination with smartphones has been Apple's rebranding of the smartphone evolution as a revolution (see Gerard Goggin's chapter in this collection). Even so, the iPhone is still to obtain market dominance in China. In a 2011 report of the Chinese smartphone market released by ZDC, the iPhone ranked fourth in Chinese smartphone brands at 7.1% after Nokia, Motorola, and HTC.

As an increasingly dominant part of Apple's empire, globally the iPhone represents a specific version of lifestyle technologies—encapsulated by the fetishization of the personal in the form of apps. This amplification of the personal occurs under a specifically Western notion of the personal, as something that is bounded to ideas of the individual and clear demarcations of the private. When reflecting upon the iPhone in non-Western contexts such as China, we can see how cultural differences play out through mobile media.

The first-generation iPhone 2G was launched in the US in mid-2007 and has been followed by various renditions each year. In China, the official introduction of the iPhone was in 2009 with the 3G model. By 2010, as a result of much marketing hype, the iPhone emerged as a desired object in China. With iPhone 4 slogans such as, "If you don't have an iPhone, well, you don't have an iPhone," the iPhone no longer represents just *a* mobile phone in China, it has come to represent *the* mobile phone, the epitome of fashionable, cool, youth cultures.[16] As one of the first examples of 4G mobile technology, many young people, especially from the *ba ling hou*, viewed the acquisition of an iPhone as an extension of their technological, mobile lifestyle.

The iPhone occupies a particular place in the hierarchies of mobile media in China.[17] Before the official launch of the iPhone 3Gs in 2009, countless 2G and 3G iPhones came into the Chinese mobile phone market via *"shui huo"*—that is, they entered China through unofficial "black market" methods. This black market situation arguably constituted the iPhone as a highly sought after and desirable object. When Apple and China Unicom began to work together with iPhone 3Gs and 4G, suddenly "official" versions could be purchased. The latest generation, the iPhone 4, was released in China on 25 September 2010. According to China Unicom, more than 200,000 Chinese customers had pre-ordered this model and there was a wild response by consumers on its release; *CNN Money*, for example, reported that, "By 8 a.m., when the doors opened at Apple's new store in Beijing's Joy City shopping mall, the queue was more than 1,000 customers long."[18] The same scene occurred when the white version of the iPhone 4 was launched on 28 April 2011; despite the fact that the white version did not feature any additional hardware upgrades, there were still thousands of people queuing outside Apple retail stores, with some even resorting to violence.[19] By May 2011, China Mobile chairman Wang Jianzhou disclosed that China Mobile had more than four million iPhone users.[20]

Thus, in China, like many other locations during its early phases of adoption and growth in popularity, the iPhone became a fashion icon. Celebrities, fashion stars, and Apple fanatics may be among the first groups to use the iPhone, yet the growing ubiquity of the iPhone has not only been in generally more affluent cities, but also in some rural areas—even Tibet. Writing on the success of the iPhone in Tibet, the *Xinhua News Agency* reports that Tibetans appear to love the iPhone[21]: "Many people, from office workers and young business executives to monks at Lhasa's major monasteries, bought the latest model, the iPhone 4, shortly after it was launched."[22]

Unquestionably, though, the main market for the iPhone is still the young and socio-economically mobile, who are synonymous with the *ba ling hou*. According to iResearch's *China iPhone User Research Report* (2011), 58.6% of users are between 25 and 34 years old; 69.7% of users have undergraduate or higher degree qualifications; and 34.4% of users have a monthly income that exceeds 5,000 RMB (AUD$770).[23] While the relatively high price of the iPhone does exclude a large portion of the market, and is thus limited to more affluent users, the price also highlights a strong correlation between social capital, economic capital, and education. Interestingly, iPhone users in China (54.5% of whom use the internet daily) are not high users of the internet when compared to other mobile phone users (71.1% of whom use the internet daily). According to iResearch, these figures reflect iPhone user's high level of engagement with its entertainment functions (i.e., games and apps) and also the specific lifestyle preferences of its main user group.

While the deployment of GPRS in China ensured that urban and rural, rich and poor people could access the internet via their mobile phones prior to the smartphone phenomenon, it will be interesting to see what the effects are for online experience that ensue from the rise of the smartphone and its countless copies, the *shanzhai*. In the case of the iPhone, according to a preliminary iResearch report, the device is viewed as both a fashion icon and "an attitude of life" (i.e., a barometer of the user's lifestyle).[24] As a tool of, and for, lifestyle, the iPhone is strongly identified with the *ba ling hou*. Within this demographic, unique types of media practices that overlay the electronic and geographic with the social and emotional can be found. This situation is especially apparent with geomedia like *Jie Pang*, as we shall see in the following case study of Chinese iPhone users.

THE iPHONE AND *JIE PANG*: A CASE STUDY

As a relatively new phenomenon, *Jie Pang* signals a new epoch in urban mobile gaming. Although the first epoch of urban mobile gaming by groups such as Blast Theory was marked by experimental reconfiguring of the urban social fabric by transforming the urban environment into a play space,[25] the latest is marked by a "gamification" of social media, whereby one tries to go to the newest, coolest places and show their friends as part of a game. While China's embrace of mobile games like *Angry Birds* and social media games like *Happy Farm* is cross-generational, *Jie Pang* is used primarily by the *ba ling hou*. In order to understand this phenomenon, we conducted online surveys with *ba ling hou* users asking them about their mobile and geosocial media practices from May to June 2011. Having initially surveyed a group of fifteen respondents, we then noticed that there was a correspondence between users having an iPhone and using the emergent LBS, *Jie Pang*, so we again conducted a series of revised surveys with six respondents. Once an activity almost solely the preserve of artists, researchers, and experimental educators, urban mobile gaming[26] has now become more mainstream with geosocial services like *Foursquare* and *Jie Pang* in China.

As a Chinese version of *Foursquare*, *Jie Pang* renders places and movement into part of its gameplay architecture. The key motivation for users is to both see where their friends are as well as to find and report on new "cool" places. While discussions in Western contexts about *Foursquare* have been concerned with ideas of privacy and surveillance (epitomized by the "please rob me" website[27]) such notions about the individual and its relationship to the social do not translate into the Chinese context. Part of the reason *Jie Pang* does not attract the same types of debates about privacy can be found in culturally specific notions like *guanxi*. As Cara Wallis observes, in the case of Beijing, the deployment of social media is closely bounded by the notion of *guanxi*.[28] Wallis notes,

> In contrast to the individual-orientated nature of western cultures, where the autonomy of the individual is presupposed, Chinese social organization has been described as relationship-orientated. In traditional Chinese culture . . . there is no unique "self" outside of social relationships and the personal obligations that inhere in those relationships . . . despite the influences of communism, industrialization, urbanization, and westernization, many have still found utility in conceptualizing the Chinese sense of self as predominantly relationally focused.[29]

For Wallis, *guanxi* is a "widely used yet ambiguous" term that can mean many things: relationships, personal connections, and social networks. The term holds similar meaning to that carried by Pierre Bourdieu's (1979) notion of "social capital" whereby knowledge is not rewarded by *what* you know but *who* you know.[30] The notion of *guanxi* is significant in the uptake of new media like *Jie Pang*. An early adopter will often persuade friends to join the new media network with the promise that it is not for everyone, but rather, just for them. Here we see that *guanxi* fosters a tightening of the close social ties that often exclude other, less close contacts—a phenomenon Ichiyo Habuchi calls "telecocooning."[31] By deploying new media like *Jie Pang*, the *guanxi* associated with users of this service can be tightened by singling those out with similar lifestyles, socio-economic, and technological backgrounds.

Jie Pang is a geosocial game almost exclusively the preserve of the *ba ling hou*. *Jie Pang* epitomizes their relationship to the internet as a space that is public and also almost compulsory. Indeed, for this generation, there is a need to be "always on"[32] and omnipresent through media, and services like *Jie Pang* facilitate this. That is to say, the *ba ling hou* use *Jie Pang* to be part of the intimate publics in which the personal practices of the everyday become, on the one hand, commodified and, on the other, further tether a sense of place (which is as much emotional and social as it is geographic and physical) with personal politics. The connection with place and location sharing is closing linked to not only narrativizing their travels for themselves, but, more importantly, as a way to share co-present intimacy with friends. Here we see that the personal is political insofar as the domestic and personal practices around these geosocial games are played out in public domains. But this public domain is a space of multiple localized communities that see their usage of *Jie Pang* as not rendering them a *phoneur*[33]—a notion Robert Luke coined to describe the user as a mere node in information and commercialization circuits—but rather, it forms a meaningful way to further enhance mobile intimacy.

While the ramifications of *Jie Pang*'s darker side—that is, the way in which such geosocial media weakens privacy and even invites stalking[34]—are yet to fully play out, its deployment does illustrate the ways in which the *ba ling hou* are part of new generations in which privacy, a concept once born out of inconvenience,[35] is now having its rules completely rewritten. Moreover, *Jie Pang* use highlights the way in which participation, as with privacy, is viewed differently in China. Unlike the West, in which such issues are viewed as a cornerstone for the rights of the individual (especially in legal terms), in China participation has a different flavor. Such notions as watching—what we call "lurking" in the West[36]—are seen as an active and positive part of participation.[37]

In the surveys we conducted, we inquired about the respondents' mobile media usage vis-à-vis their broader media practices. After conducting two series of surveys with ten people from the *ba ling hou*, it became apparent that there was a correlation between iPhone users and interest in new media applications like the LBS *Jie Pang*, perhaps because those that can financially afford an iPhone could also afford the significant temporal commitments that are demanded of such playful but time-intensive leisure activities. Given this curious correlation between iPhone users and *Jie Pang*, we decided to investigate further—especially the different ways that presence is being played out in relation to usage of these new media services—by conducting additional surveys that, this time, were focused upon combined iPhone and *Jie Pang* use. While we accept that not all iPhone users play *Jie Pang* (and vice versa), in the second set of surveys, we were interested in this combined use, especially in light of current movements toward a mobile-scape that is increasingly dominated by geomedia or LBS services.[38] The questions asked of survey participants sought to explore the motivations, expectations, possibilities, and limitations of their use of *Jie Pang*. Given that one of the authors of this chapter, Kay Gu, is an iPhone and *Jie Pang* user, potential survey respondents were primarily recruited via snowballing from her network connections through SNSs. The demographic sought included the *ba ling hou*—respondents born between 1980 and 1989. We sought to get an equal amount of female to male respondents. In the first survey, we had equal gender balance; however, in the iPhone survey we had two females and four male respondents. As aforementioned, the first series of surveys about smartphones had ten respondents, which was followed by a revised iPhone-focused survey to six respondents.

What emerged from the surveys was that, for many, the main motivation for *Jie Pang* use was to record the places they went to, for both themselves (i.e., as an *aide-mémoire*) and also in order to share this information with their friends as a kind of "networked memorialization." According to Ai (female, aged 25),[39] *Jie Pang* use was in order "to record where I go to everyday. It's like a diary with location." The other respondents shared Ai's sentiment—many viewed recording locational information via *Jie Pang* as both for their own and others' benefit. For Bai (male, aged 27), *Jie Pang* was primarily used to "record where I had been" and to have this information "synchronized to social networks such as Weibo to share with my followers." For others, they did not record everywhere they went, but rather, used *Jie Pang* only for those occasions when they went to new places.

Ai observed more females than males using *Jie Pang*, especially in terms of everyday practices that were linked with that of camera phone picture taking. She put this down to the fact that, in her opinion, "guys usually don't want to share or record" this sort of mundane, everyday information. When asked if she worried about people knowing where she was, given that issues of surveillance are particularly important for female users, Ai responded, "yes, to some extent. So that's why I choose [carefully] which SNS platform to synchronize with my check in." For this respondent, like the other female respondents surveyed, she uses *Jie Pang* "every new place I go to every day" and in order "to note my footprint." Although she began using *Jie Pang* as a novelty to play with friends and "to win the virtual badge" (a prize for visiting certain places which is part of the designed game play), as was the case with others surveyed, her motivation soon changed to one that carried far greater significance as part of her everyday diarization and narrations of places that interwove her geographic and electronic footprints.

For Bai, *Jie Pang* was important in recording and archiving his activities and journeys. Unlike Ai, Bai did not record each place he went to everyday, just a few highlights. Here we see the way in which gender inflects the ways in which *Jie Pang* relates to ongoing endeavors to narrate the everyday. Bai viewed *Jie Pang* as a tool for showing where he was when he wanted people to know. Sometimes checking in on *Jie Pang* was accompanied by taking pictures of the place, an activity Bai definitely viewed as gendered, stating, "Usually females would spend more time on it than men, and take more photos." Both these informants noted that recently there had been a growth in different LBSs on smartphone and PC platforms, and that groups of friends would use similar ones—in short, the deployment of an LBS reflected the *guanxi*.

For Chung (male, aged 30), *Jie Pang* represented a new suite of mobile applications that prompted users to think about place in different ways. For this respondent, no clear difference in gender usage was noted. Rather, the main motivation for using *Jie Pang* was to show off a new place. Unlike Ai and Bai, for whom *Jie Pang* had moved beyond the honeymoon stage and had become an important everyday practice in narrating a sense of place and mobility, Chung still viewed it as a novelty. Similarly, for Huang Fu (male, aged 26), *Jie Pang* afforded him a playful and inventive way to show his current location to his friends.

For Baozhai (female, aged 26), like the other respondents, *Jie Pang* provided an innovative way to narrate one's journey that combined a sense of place and networked sociality. She noted that surveillance was an issue and this led her to sometimes not use the service. Even so, Baozhai would use *Jie Pang* in a variety of locations, including "at the airport to track my flight records" and "when I go to some supreme restaurants or hotels which I rarely go to." When asked about whether she could imagine ever checking in at every place that she went to, she replied that such a feat would be "impossible." Moreover, "some routine places like home or company are not worthy of checking in." Here, the choice of words "not worthy" is important. Asked about the future of media and where she saw *Jie Pang*, if at all, Baozhai stated:

> It's absolutely a fad, just like microblogs overtaking blogs. What's more, *Jie Pang* is not meaningful for people like reading blogs, which has more significant content. DianPing [another LBS] could totally overtake *Jie Pang* if it used all the concepts of *Jie Pang*'s LBS game.

Jianjun (male, aged 27) used *Jie Pang* to "check in at some cool places and share [these] with my friends." This was especially so if he was to go to a "new cool place, a meaningful place [as well as] some places with medals." As a built-in component of game play, the activity of winning medals— achieved by going to places and recruiting others—rated relatively low in terms of motivations for use. The playful elements were actually created by users outside the intended game play components—play happened when users were able to connect (both in online and sometimes offline spaces) and comment. The use of *Jie Pang* via SNSs meant that users knew only their friends could see and comment upon where they were. Unperturbed by worries about surveillance or privacy, Jianjun viewed *Jie Pang*'s role as essentially a way of sharing and being co-present with friends. One of the ways to create an intimate public—that is, so only one's friends could view *Jie Pang* check-ins—was if they checked in through an SNS. This technique was deployed by most of the respondents as a way to foster some form of privacy and to make sure that the overlay between geographic and electronic co-presence was done within the context of friends or *guanxi*. Here we see how *Jie Pang* brings together co-presence, place, and mobility in ways that infuse them with social capital significance. As Jianjun puts it, "When I choose to use *Jie Pang* that means I would love to share these [places and experiences] with my friends." Jianjun noted that most of his friends were using *Jie Pang*: "In my opinion, *Jie Pang* is a tool for fun and interaction with friends, since you can connect your *Jie Pang* account with Kaixin, Ren-ren, and Sina microblog (only my friends can see my check-ins)."

Beyond the hype of *Jie Pang* as a first-generation commercial mobile LBS game, already we are starting to see some of the ways in which such geosocial applications are being used by respondents as a meaningful part of everyday practice, in which the electronic is overlaid with the social, emotional and geographic in ways that both rehearse older practices of socio-spatial connectivity at the same time as they create new social geographies. Particularly interesting is the way that these cartographies are often gendered in their deployment, with males and females using *Jie Pang* in different ways to map their socio-emotional practices onto everyday coordinates. While being informed by factors such as gender, age, and class, geomedia such as *Jie Pang* also signal emergent types of connected presence and ambient intimacy. By overlaying the socio-emotional with the geographic and electronic, geomedia suggests the value of revisiting and perhaps reconceptualizing theories that are concerned with the interactions and intersections between presence, co-presence, and intimacy.

CONNECTED PRESENCE AND AMBIENT INTIMACY

There has been longstanding critical interest in the concept of presence, which spans a range of diverse areas including everything from telemedicine, flight simulation training, immersive virtual environments and video conferencing, to social networking services, electronic discussion lists, computer-mediated communications, and mobile media.[40] Broadly defined, in the present context, presence can be understood as referring to:

> [T]he degree to which geographically dispersed agents experience a sense of physical and/or psychological proximity through the use of particular communication technologies.[41]

In early internet scholarship on presence, there was considerable semantic debate, specifically around attempts to differentiate between technologically mediated and unmediated forms of presence.[42] For a number of critics participating in these debates, any such distinction was untenable; this is because, as Giuseppe Mantovani and Giuseppe Riva have noted, it fails to acknowledge that "presence is always mediated" and that it is culturally constructed.[43] And yet, as important as this qualification is, equally significant is recognizing that "the ability of the subject to elide or ignore this mediation is crucial to the presence effect."[44] It is in this way that,

> Presence can be understood as a psychological state in which the person's subjective experience is created by some form of media technology with little awareness of the manner in which technology shapes this perception.[45]

These insights are developed further in an important study of presence and communications technologies (both old and new) by Esther Milne. One of her key insights is that, paradoxically, face-to-face presence involves a type of absence (or "disembodiment") that is the product of distance in space and time. Distance can occur on many levels—memory, emotions, gestures, and language—yet can nevertheless achieve a sense of intimacy that is the result of "the eclipse of the material medium that supports" communication, as well as the "temporal or physical [that is, geographic] obstacles" that stand between those communicating.[46] Intimacy as it is established between correspondents is both dependent on and prompted by "their temporal and spatial distance from one another."[47]

Strong interest in questions of presence has continued in scholarship that examines the cultural uses of mobile media. For instance, Hjorth builds on Milne's interest in earlier communications technologies and forms of "postal presence" to argue that "SMS re-enacts 19th Century letter-writing traditions," and, in this way, "the intimate co-presence enacted by mobile technologies should be viewed as part of a lineage of technologies of propinquity."[48]

Somewhat differently, in a detailed study of presence in relation to fixed line, mobile, and SMS services, Christian Licoppe remarks that "the ways absent ones make themselves present have been many" and the "material resources that support presence [are] growing in number with the advent of the so-called 'information society'" to such an extent that "co-present interactions and mediated communication seem woven in a seamless web."[49] Licoppe's broader point is that mobile technologies are being used not just to "compensate for the absence of our close ones," but rather "are exploited to provide a continuous pattern of mediated interactions that combine into 'connected relationships,' in which the boundaries between absence and presence eventually get blurred."[50]

For Kenneth Gergen, meanwhile, our experience of mobiles is characterized by a rather different form of "absent presence" that is constituted by two competing movements. On the one hand, and echoing Licoppe's point above, there is the issue of how "the cell phone facilitates new integrations of the absent and the present in more subtle ways."[51] On the other hand, this comes at a cost: when mobile phone users are "locked in cell phone conversation they cease to be full participants in the immediate context" in which they are situated.[52] Gergen's concern is that, increasingly, "one is physically present but is absorbed by a technologically mediated world of elsewhere."[53] Thus, the challenge, as Gergen sees it, is how to negotiate the tensions between these two senses of "absent presence."

Presence is also figured in the mobile literature in other ways that are developed from close examinations of situated, bodily encounters with mobiles, place, and communicative exchange. This work tends to speak less to Gergen's concerns than they do to Licoppe's point above about the blurring of presence and absence. For instance, in her important study of youth mobile use in Tokyo, Mizuko Ito describes how, even "after young people have converged in physical space" to meet face-to-face, "mobile communication does not necessarily end."[54] Ito coins the term "augmented co-presence" as a way of making sense of these complex absent/present communicative interactions, and how they can lead to "new senses of place being constructed as a hybrid between co-located and remote social contact."[55]

In addition, there is the work of Ingrid Richardson that combines a concern for "techno-somatics" and post-phenomenology in her explorations of the "complex and dynamic range of place interactions and differing modalities of presence"[56] that characterize the micro-scale of our embodied engagements with mobile media devices. For Richardson, central to these engagements is the issue of distraction—that is, "how our attention becomes divided when we speak on the phone, send or receive a text message, or play a game on the mobile."[57] This occurs, she argues, in ways that involve a canny and subtle form of "environmental knowing"[58] that is attuned to both the specific requirements of mobile game play while retaining a "crucial peripheral awareness of one's spatial surroundings."[59] Thus, while recognizing the concerns expressed by Gergen above, Richardson argues that, at a bodily level, intricate interactions are at play that need to be acknowledged: "The 'sensing' of mobile communication and interactive media elicits an intimately audio, visual, sometimes haptic, 'handy' and visceral awareness, a mode of embodiment which demands the ontological coincidence of distance and closeness, presence and telepresence, actual and virtual."[60]

For the *ba ling hou* generation, especially those who have travelled to Shanghai for the purpose of study, their use of mobile communications technologies fits squarely within the kinds of understandings of presence described above (especially as detailed by Licoppe). For instance, with this particular demographic, mobile phones compensate for and work to overcome the geographical distance that separates them from home and from their parents. As Hjorth writes elsewhere, "Through the constant contact afforded by mobile connectivity, the *ba ling hou* are free to roam with their 'parents-in-the-pocket' . . . In this way, mobile and social media create a relocation of these domestic ties outside the physical space of the home."[61] Moreover, through repeated, habitual use, their engagement with media services such as QQ "allows for the small, informal, micro exchanges like those experienced when one is in intimate proximity." To return to Licoppe's earlier phrase, connected presence via mobile communication works to "provide a continuous pattern of mediated interactions that combine into 'connected relationships,' in which the boundaries between absence and presence eventually get blurred."[62] What emerges, as a way of ameliorating the effects of geographical distance and "absence," are forms of presence negotiated through "an affective cartography that is marked and maintained by the ephemeral, personal and micronarratives of social mobile media."[63]

Our respondents' use of the geomedia service *Jie Pang*, however, involves a quite different set of engagements with presence. In contrast to the above, the idea of presence evoked by their use of *Jie Pang* is closer to that described by Okabe in his study of camera phone usage, in that it engages users in "creating a sense of presence in other people's lives without needing to talk or be physically present."[64] The forms of connected presence that result are described by Leisa Reichelt as "ambient intimacy"—that is, "being able to keep in touch with people with a level of regularity and intimacy that [one] wouldn't usually have access to, because time and space conspire to make it impossible."[65] A key use of *Jie Pang*, then, is in maintaining presence within social networks via regular locational postings. And, yet, our survey respondents indicate that the construction and maintenance of ambient intimacy, while crucial, is only one facet of their engagements with *Jie Pang*. Another key facet is the social capital implications of *Jie Pang* use.

Following Nan Lin's definition, social capital can here be understood as "resources embedded in social relations and social structure, which can be mobilized when an actor wishes to increase the likelihood of success in a purposive action."[66] As noted elsewhere, this formulation contains two important components: first, resources are embedded in social relations rather than individuals; second, access to and use of these resources reside with individual actors.[67] "Capital," in this context, is viewed by Lin as a "social asset by virtue of actors' connections and access to resources in the network or group in which they are a member."[68] In the context of the present discussion of the *ba ling hou*, access to resources involves the technological knowledge acquired during their education, which is a prized social asset within the context of their extended familial network. Importantly, it also involves access to technological resources, such as the highly symbolically charged signifier of affluence and "coolness" that is the focus of this collection: the Apple iPhone.

Finally, and more specific to *Jie Pang*, the social asset at play here is also geographical or locational knowledge. Echoing Ilkka Arminen's point that when it comes to SNSs, social context rather than "pure geographical location" is generally of greatest user interest,[69] what seems to be most at stake in *Jie Pang* is new knowledge about particular sites and what these are likely to signify within social network settings. Thus, many of those surveyed revealed that the higher the perceived level of novelty or uniqueness that is seen to be associated with a place (such as, in the words of one respondent, a "supreme" restaurant or hotel), the higher the likelihood that their presence in and knowledge of this place will be recorded via *Jie Pang*, as "routine places like home or [their] company are not worth checking in." In this way, it is geographical (or "environmental") "knowing" and an appreciation of the "capital" that is invested in and carried with this knowledge that is paramount, and which forms a vital resource within the participants' wider peer network.

CONCLUSION: PRESENCE AND GEOSPATIAL SOCIALITY

Geosocial mobile media services like *Foursquare* render *sociality* and *place* into a networked game. In the case of the Chinese equivalent, *Jie Pang*, this geosocial game was almost exclusively taken up by the *ba ling hou*. It epitomized their relationship to the internet as a space that is public and also almost compulsory. Indeed, for this generation, there is a need to be "always on" and omnipresent through media like *Jie Pang*. Moreover, though not limited to iPhone users, the uptake and early adoption of *Jie Pang* has been preliminarily linked to a type of "iPhone attitude." As identified in the case study, users viewed the typical iPhone user as fashionable, trendy and fun, characteristics they also applied to *Jie Pang*.

A service such as *Jie Pang* enables the *ba ling hou* to be part of intimate publics in which the personal practices of the everyday become, on the one hand, commodified and, on the other hand, further tethered to a sense of place that is as much emotional and social as geographic and physical. Social relations are thus played out in public domains in ways that foreground both networked social and place-based settings as they are negotiated in combination. Through *Jie Pang*, the public domain becomes a space of multiple localized communities whose members are not wandering *phoneurs* or nodes in information and commercialization circuits, but are tangling up place and sociality to construct and maintain mobile intimacy.

As we have suggested, the deployment of geomedia like *Jie Pang* illustrates the need to revise the relationship between presence, intimacy, and place. Drawing on earlier concepts of presence we have suggested that the form of connected presence most afforded by *Jie Pang* is that of "ambient intimacy" in which a sense of shared co-presence is produced through social media that is overlain with a sense of place and maintained through regular locational postings. However, another key factor in maintaining ambient intimacy is a culturally specific notion of social capital in the form of *guanxi*. The saliency of *guanxi* also highlights that while geomedia is creating new forms of ambient intimacy it is also rehearsing and tightening older social ties.

NOTES

1. Daniel Miller, *Tales from Facebook* (Cambridge, UK: Polity, 2011), 212.
2. Ibid., 204.
3. Jack Linchuan Qiu, *Working-Class Network Society: Communication Technology and the Information Have-Less in Urban China* (Cambridge, MA: MIT Press, 2009).
4. Jack Linchuan Qiu, "Wireless working-class ICTs and the Chinese informational city," *The Journal of Urban Technology* 15(3), 2008: 57–77; Stephanie Hemelryk Donald, Michael Keane, and Yin Hong, eds., *Media in China: Consumption, Content and Crisis* (London: Routledge, 2002).
5. Birdabroad, http://birdabroad.wordpress.com/ 23 July 2011 (accessed); an example of the story in media can be found at: "Editorial, fake Apple stores found in Kunming City, China," 23 July 2011, *BBC news*, http://www.bbc.co.uk/news/technology-14236786?print=true
6. Richard Robison and David S. G. Goodman, eds., *The New Rich in Asia: Mobile Phones, McDonalds and Middle-class Revolution* (London: Routledge, 1996); Cara Wallis, "(Im)mobile mobility: marginal youth and mobile phones in Beijing," in Rich Ling and Scott W. Campbell, eds., *Mobile*

Communication: Bringing Us Together and Tearing Us Apart (New Brunswick: Transaction books, 2011), 61–81.

7. Jens Damm and Simona Thomas, eds., *Chinese Cyberspaces Technological Changes and Political Effects* (New York: Routledge, 2006); Christopher R. Hughes and Gudrun Wacker, eds., *China and the Internet: Politics of the Digital Leap Forward* (London: Routledge Curzon, 2003); Zixue Tai, *The Internet in China: Cyberspace and Civil Society* (New York: Routledge, 2006).

8. Pui-Lam Law and Yinni Peng, "The use of mobile phones among migrant workers in Southern China," in Pui-Lam Law, Leopoldina Fortunati and Shanhua Yang, eds., *New Technologies in Global Societies* (Singapore: World Scientific, 2006), 245–258; Qiu, *Working-Class Network Society*; Qiu, "Wireless working-class ICTs"; Wallis, "(Im)mobile Mobility."

9. Axel Bruns and Joanne Jacobs, eds., *Uses of Blogs* (New York: Peter Lang, 2006); Geert Lovink, *Zero Comments: Blogging and Critical Internet Culture* (New York: Routledge, 2007); Haiqing Yu, "Blogging everyday life in Chinese internet culture," *Asian Studies Review* 31(4), 2007: 423–433.

10. Pamela T. Koch, Bradley J. Koch, Kun Huang, and Wei Chen, "Beauty is in the eye of the QQ user: instant messaging in China," in Gerard Goggin and Mark McLelland, eds., *Internationalizing Internet Studies* (London: Routledge, 2009), 265–284; Yu, "Blogging everyday life."

11. CNNIC [China Internet Network Information Center] *Statistical Survey Report on the Internet Development in China, 2009* (January 2009), http://www.cnnic.cn/uploadfiles/pdf/2009/3/23/131303.pdf (accessed 14 June 2009).

12. Ibid.

13. Jun Wang and Stephen Siu Yu Lau, "Gentrification and Shanghai's new middle-class: another reflection on the cultural consumption thesis," *Cities* 26(2), 2009: 57–66.

14. Larissa Hjorth and Michael Arnold, "Home and away: a case study of emerging technocultures and mobility in Shanghai," in Patrick L Law, ed., *New Connectivities in China: Virtual, Actual, and Local Interactions* (Dordrecht, The Netherlands: Springer, 2011).

15. CNNIC, *Statistical Survey Report.*

16. IDC [International Data Centre], "The three trends of Chinese smartphone market," 2011, http://www.idc.com/about/viewpressrelease.jsp?containerId=prUS22689111 (accessed April 2011).

17. Li Qiaoyi, "New iPhone 4 creates battle for sales," *Global Times,* 26 September 2010, http://business.globaltimes.cn/industries/2010-09/577062.html (accessed 20 January 2011).

18. Philip Elmer-DeWitt, "Massive crowds greet iPhone 4 in China," *CNNMoney* (2010). http://tech.fortune.cnn.com/2010/09/25/massive-crowds-greet-iphone-4-in-china/ (accessed 25 May 2010).

19. Editorial, "White iPhone 4 launch turns violent at Beijing store," *PCMagazine,* 2011, http://www.pcmag.com/article2/0,2817,2385067,00.asp (accessed 9 May 2011).

20. Editorial, "China Mobile has 4 mln iPhone Users," *SINA Tech,* 2011, http://www.marbridgeconsulting.com/marbridgedaily/2011-05-19/article/46321/china_mobile_has_4_mln_iphone_users (accessed 19 May 2011).

21. Editorial, "Chinese state media: Tibetans love the iPhone," *Xinhua News Agency,* 24 May 2011 (accessed 19 May 2011).

22. Editorial, "China Mobile."

23. iResearch "China iPhone user research report," iResearch, 2011 http://www.iresearchchina.com/view.aspx?id=9136 (accessed May 2011).

24. Ibid.

25. Adriana de Souza e Silva and Larissa Hjorth, "Urban spaces as playful spaces: a historical approach to mobile urban games," *Simulation and Gaming* 40(5), 2009: 602–625.
26. Ibid.
27. Websites such as www.pleaserobme.com have been made to "raise awareness about over-sharing." Pleaserobme claims that increased locational awareness is creating a loss of locational privacy.
28. Wallis, "(Im)mobile mobility."
29. Wallis, "(Im)mobile mobility," 67.
30. Pierre Bourdieu, *Distinction: A Social Critique of the Judgment of Taste*, trans. R. Nice (Cambridge, MA: Harvard University Press, 1984).
31. Ichiyo Habuchi, "Accelerated reflexivity," in Mizuko Ito, Daisuke Okabe and Misa Matsuda eds., *Personal, Portable, Pedestrian: Mobile Phones in Japanese Life* (Cambridge, MA: MIT Press, 2005), 165–182.
32. Naomi Baron, *Always On* (Oxford: Oxford University Press, 2009).
33. Robert Luke, "The phoneur: mobile commerce and the digital pedagogies of the wireless web," in Peter Pericles, ed., *Communities of Difference: Culture, Language, Technology*, Trifonas (Palgrave: London, 2006), 185–204.
34. Iain Sinclair, *Lights Out for the Territory* (London: Penguin Books, 1997); Alison Gazzard, "Location, location, location: collecting space and place in mobile media," *Convergence: The International Journal of Research into New Media Technologies* 17 (4), 2011: 405–417.
35. Clay Shirky, "Here comes everybody," presented at *the Aspen Ideas Festival*, 30 June to 8 July 2008, Aspen, Colorado, http://fora.tv/2008/07/06/Clay_Shirky_on_Social_Networks_like_Facebook_and_MySpace#chapter_01 (accessed 20 January 2009).
36. Kate Crawford, "Following you: disciplines of listening in social media," *Continuum* 23(4), 2009: 523–535.
37. Gerard Goggin and Larissa Hjorth, "Waiting to participate: introduction," *Communication, Politics & Culture* 42(2), 2009: 1–5.
38. For a diagrammatic representation of the current geosocial mediascape as accessed via mobile phones, see http://geosocialmedia.org/ (accessed 2 July 2011).
39. All names used are pseudonyms.
40. Esther Milne, *Letters, Postcards, Email: Technologies of Presence* (New York: Routledge, 2010), 165.
41. Ibid.
42. Ibid., 165–166.
43. Giuseppe Mantovani and Giuseppe Riva, "'Real' presence: how different ontologies generate different criteria for presence, telepresence and virtual presence," *Presence: Teleoperators and Virtual Environments* 1(1), 1998: 540–550.
44. Milne, op. cit., 166.
45. Ron Tamborini and Paul Skalski, cited in Milne, *Letters, Postcards, Email*, 2010, 166.
46. Milne, *Letters, Postcards, Email*, 2010, 9.
47. Ibid., 15. It is in this way that Milne counters arguments expressed by Andrew Feenberg that physical co-presence constitutes the "ideal paradigm" of communicative intimacy, by provocatively suggesting that, "all intimacy is perhaps always 'virtual'." Ibid., 16.
48. Larissa Hjorth and Ingrid Richardson, "The waiting game: complicating notions of (tele)presence and gendered distraction in casual mobile gaming," *Australian Journal of Communication* 36(1), 2009: 26. See also, Larissa

Hjorth, "Locating mobility: practices of co-presence and the persistence of the postal metaphor in SMS/MMS mobile phone customization in Melbourne," *Fibreculture Journal* 6, 2005, http://six.fibreculturejournal.org/fcj-035-lo-cating-mobility-practices-of-co-presence-and-the-persistence-of-the-postal-metaphor-in-sms-mms-mobile-phone-customization-in-melbourne/ ; Larissa Hjorth, "Postal presence: a case study of mobile customisation and gender in Melbourne," in Peter Glotz and Stefan Bertschi, eds., *Thumb Culture: Social Trends and Mobile Phone Use* (Berlin, Germany: Bielefeld, 2005), 53–66; Esther Milne, "'Magic bits of paste-board': texting in the nineteenth century," *M/C Journal* 7(1), 2004, http://journal.media-culture.org.au/0401/02-milne.php (accessed 4 June 2006).

49. Christian Licoppe, "'Connected' presence: the emergence of a new repertoire for managing social relationships in a changing communication technoscape," *Environment and Planning Design: Society and Space* 22(1), 2004: 135–156.

50. Ibid., 135–136.

51. Kenneth Gergen, "The challenge of absent presence," in James E. Katz and Mark Aakhus, eds., *Perpetual Contact: Mobile Communication, Private Talk, Public Performance* (Cambridge: Cambridge University Press, 2002), 239.

52. Ibid., 239.

53. Ibid., 227.

54. Mizuko Ito, "Mobiles and the appropriation of place," *Receiver* 8, 2003, http://academic.evergreen.edu/curricular/evs/readings/itoShort.pdf (accessed 10 October 2011).

55. Ibid.

56. Ingrid Richardson and Rowan Wilken, "Parerga of the third screen: mobile media, place and presence,", in Rowan Wilken and Gerard Goggin, eds., *Mobile Technology and Place* (New York: Routledge, 2012, forthcoming).

57. Larissa Hjorth and Ingrid Richardson, "The Waiting Game," 30.

58. Ibid., 33.

59. Ibid., 32.

60. Ingrid Richardson, "Mobile technosoma: some phenomenological reflections on itinerant media devices," *Fibreculture Journal* 6, 2005, http://six.fibre-culturejournal.org/fcj-032-mobile-technosoma-some-phenomenological-reflections-on-itinerant-media-devices/; see also, Ingrid Richardson and Rowan Wilken, "Haptic vision, footwork, place-making: a peripatetic phenomenology of the mobile phone pedestrian," *Second Nature: International Journal of Creative Media* 1(2), 2009, http://secondnature.rmit.edu.au/index.php/2ndnature/article/view/121/35.

61. Larissa Hjorth, "Still mobile: a case study on mobility, home and being away in Shanghai," in Rowan Wilken and Gerard Goggin, eds., *Mobile Technology and Place* (New York: Routledge, 2012, forthcoming).

62. Licoppe, "'Connected' Presence," 136.

63. Hjorth, "Still mobile."

64. Daisuke Okabe, "Emergent social practices, situations and relations through everyday camera phone use," *Mobile Communication and Social Change: International Conference on Mobile Communication*, Seoul, Korea, 18–19 October 2004, http://www.itofisher.com/mito/archives/okabe_seoul.pdf (accessed 30 June 2009).

65. Leisa Reichelt, "Ambient intimacy," post to disambiguity blog, March 1 2007, http://www.disambiguity.com/ambient-intimacy/ (accessed 10 January 2011).

66. Nan Lin, *Social Capital: A Theory of Social Structure and Action* (New York: Cambridge University Press, 2001), 24.
67. See Rowan Wilken, "Bonds and bridges: mobile phones and social capital debates," in Rich Ling and Scott W. Campbell, eds., *Mobile Communication: Bringing Us Together and Tearing Us Apart* (New Brunswick, Transaction Publishers, 2011), 127–149.
68. Lin, op. cit, 24.
69. Ilkka Arminen, "Social functions of location in mobile telephony," *Personal and Ubiquitous Computing* 10(5), July 2006: 319–323.

5 "In Bed with the iPhone"
The iPhone and Hypersociality in Korea

Dong-Hoo Lee

THE iPHONE SHOCK

Before the launch of Apple's iPhone in Korea in November 2009, the domestic market share of local handset makers such as Samsung Electronics and LG Electronics had reached almost 90 percent, and the number of smartphone subscribers accounted for merely 2.6 percent of the population between the ages of 12 and 59.[1] After its launch, the number of iPhone users reached 500,000 in four months. Korea was the 85th country in the world to introduce the iPhone, but it became one of the top seven countries where iPhones sold more than 500,000 within a year.[2] Subsequently, there was a sharp increase in smartphone subscriptions in Korea. According to the Korea Communications Commission, the number of smartphone users rose from 800,000 in December 2009, to 7.22 million in December 2010 and to 10.02 million in March 2011.[3] One in five Koreans uses a smartphone, and the number of iPhone users accounts for 27 percent of total smartphone users.

The impact of the iPhone on Korean society was called the "iPhone shock" because it influenced not only the industrial practices of both Korean telecommunication companies and local handset makers, but also mobile phone use. Korean telecommunication companies used to disable wireless internet functionality that would threaten their revenue from data services and only allowed users to upload and download data through their built-in programs. Such policies of downsizing specifications mostly limited mobile internet uses to the consumption of ring tones and wallpaper for mobile phones.[4] The iPhone was most welcome in Korea since it provided mobile internet and open-ended application services as its basic functions. After the successful rollout of the iPhone, Korean telecommunication companies raced to expand the infrastructure for Wi-Fi and introduced flat-rate plans for unlimited data usage so users could access mobile internet without worrying too much about their bills. Meanwhile, domestic handset makers, who once competed for mobile phone technologies to advance image quality, videophone function and digital multimedia broadcasting (DMB), have turned their attention to smartphones that can easily handle wireless internet services and various application programs.[5] Korean consumers began to demand from domestic handset makers the same convenience and quality as the iPhone.

Approximately a year after the introduction of the iPhone, most Korean newspapers ran a special feature on the iPhone's impact. As seen in their titles, for instance, "The iPhone's First Anniversary, It Changed Korea!,"[6] "The Ongoing 'Smart Revolution'—A Year after the iPhone's Landing in Korea"[7] and "A Year after the Introduction of the iPhone, 'Mobile Big Bang',"[8] the Phone was depicted as a revolutionary agent that created the smartphone boom in Korea and brought about enormous changes in the policies of telecommunication companies, handset makers and internet portal companies. Also, since early 2010, more than a hundred books were published covering the "iPhone effect." For instance, Kim's *Apple Shock* describes the situation one month after the iPhone's introduction as follows,

> One week after its release, there were shifts in mobile phone market and IT industry in Korea, and people began to fuss over "what the iPhone and Apple are." The scope of Apple shock driven by the iPhone was expanding. The iPhone became a symbol for innovation, and trend-setting fields such as fashion tried to borrow the iPhone's image. Business culture was rapidly transformed from the analog, to the PC-based digital, and to "Smart Biz."[9]

Popular discourse in Korea showed a high interest in the innovative characteristics of the iPhone and the creative forces of Steve Jobs and Apple. *The Secret of Apple's Success Story*,[10] *Strategies of the iPhone and the iPad*[11] and *Steve Jobs' Creative Charisma*[12] were a few examples that analyzed and discussed the success of Apple and Steve Jobs. There were also attempts to critically assess the Korean IT industry after the iPhone, such as *The Decline of the Korean IT Industry*[13] and *The Unsmart Smart Wars*.[14] In these discourses, the iPhone was received as an icon of innovation, a technology trendsetter, and as a medium for the so-called smartlife.

This chapter examines changes in the modes of mobile communication sparked by the iPhone. That is, it considers how Korean people's "personal communication systems" have been reconstructed with the use of iPhones. Jeffrey Boase uses the notion of a personal communication system to examine how individuals use multiple types of communication media in combined ways to stay connected to their personal networks.[15] This idea avoids the simple causal perspective that singles out the technological specificity of a particular medium and its direct effect and helps us to understand the impact of a new communication technology in a holistic way. When a new communication technology becomes a part of the everyday media environment, one's personal communication system is likely to be altered due to the changing social significance of existing communication technologies as well as new ones. People experience different inter-media relations as well as intra-media relations in their shifting media environments and combine old and new media differently. That is, the iPhone may affect the usage of existing media such as fixed internet, newspaper, television and personal mobile devices and also transform the ways of using multimedia functions through the mobile phone. When people become accustomed to the iPhone, they are in a new media environment and have a new personal communication system where existing communication practices co-exist with new ones.

One of the most notable phenomena that paralleled the iPhone boom in Korea was a significant increase in the number of Korean subscribers to social network sites such as Twitter and Facebook, which can be accessed conveniently in real time via smartphones. From September 2009 to September 2010, the number of Korean visitors to Twitter and Facebook increased 528 percent and 650 percent, respectively.[16] There was a sharp increase in the uses of mobile instant messengers such as *WhatsApp* and *KakaoTalk*. *KakaoTalk* was the most downloaded application for the iPhone in Korea in 2010, and it became an essential application for all smartphone users in Korea. Only a year or so after its launch, nine in ten smartphone users in Korea subscribed to *KakaoTalk*.[17] When the iPhone—working as a convergent medium that remediates various social communication media including phone calls, SMS (short messaging services), instant messengers, e-mails, and SNSs (social networking systems)—is used in everyday life, each social communication medium within and in connection with the iPhone tends to have new social significance, reconstituting personal communication environments. The iPhone allows people to exist in a virtual mobile social space with multi-layered communication contexts and to have a new mode of mobile connectivity. This chapter examines how the iPhone refashions the various affordances of social media in terms of Korean mobile media practices. Firstly, I discuss smartphones as a convergent hyper-medium and how smartphones are shaping and being shaped by earlier mobile communication and internet contexts. Then the chapter turns specifically to a case study of twenty-something iPhone users, the age-group most active in adapting and using the iPhone for sociality. Based on in-depth interview data collected from 31 Korean iPhone users in their twenties, I discuss how the iPhone has effected personal communication systems and mobile social communication practices and led to a new sociability.

SMARTPHONES AS A CONVERGENT HYPER-MEDIUM

Mobile phones allow people to have conversations with remote contacts any time and anywhere. With these portable devices, people can contact their communication partners immediately, regardless of the distance that physically separates them. According to James Katz and Mark Aakhus, such mobile connectivity makes it possible to keep "perpetual contact" so that people continue to have conversations via voice communication or text messaging on the move.[18] Mizuko Ito and Daisuke Okabe emphasize that text-based mobile messaging enables people to maintain an "ongoing background awareness of others" and, thus, to have a sense of "ambient virtual co-presence" as if they shared a social space and could have a conversation with each other whenever they wanted.[19] Christian Licoppe also writes that the mobile phone's mediation of social interactions at a distance can create a "connected presence"[20] where communication partners are always available. The experiential timeframe afforded by mobile phones develops such a sense of connectedness or co-presence. Ruth Rettie suggests that when people make calls, they expect to share a timeframe and maintain "continuous focused attention" during mediated conversation.[21] The synchronicity of mobile phone calls enables people to have a shared focus of attention and to have a sense of being together without face-to-face meetings. Rettie also argues that even though SMS is technically an asynchronous medium where transmission and reception cannot occur simultaneously, it tends to be perceived as "near-synchronous" because "most people carried their mobile phones with them and were expected to read their messages almost immediately."[22] Mobile phone connectivity, accessibility and portability make it possible to communicate immediately, to have a collaborative interaction without physically meeting and to have a sense of being connected regardless of time and location.

Although the mobile phone enables people to enlarge the scope of their social interactions beyond their immediate social surroundings, it is usually used as a tool to intensify social interactions within existing strong ties and to strengthen close relationships with familiar others.[23] Through mobile communication and its continuous connectedness, social bonds and friendships with intimates can be reinforced. Normally, sharing focused interactions with communication partners at a distance would be distracting in real co-located or face-to-face meetings. That is, people are more likely to be dislocated from their immediate surroundings and to reduce their involvement in a physically co-present context; people pursue a sense of security by building their own "telecocoons" that, as Ichiyo Habuchi writes, work as "a zone of intimacy in which people can continuously maintain their relationships with others whom they have already encountered without being restricted by geography and time."[24] On the one hand, mobile phones allow people to move into a mediated social space to interact with intimates and acquaintances in different locales and thereby maintain strong social bonds with them. On the other hand, people can select those with whom they speak and share intimacy with by their mobile phone use and avoid insecure contacts with strangers in urban public settings. An individual may be physically present in one location but virtually present elsewhere. Kenneth Gergen describes such a state as "absent presence." He argues that the form of absent presence fostered by mobile phones is "inimical to community, relations in depth, the sense of self, moral character and functional linkages between realms of meaning and action" because people are freely dissociated from their immediate and practical surroundings.[25] This history of mobile communication is further expanded and also transformed as mobile media moves toward smartphones. In the case of the iPhone, we can examine how its technological affordances are shaping and being shaped by existing mobile communication practices, and in turn how these transformations are affecting personal communication systems more broadly.

The iPhone is a particular version of the smartphone—part of the Apple empire (see Jean Burgess's and Gerard Goggin's chapters in this collection.) In this tradition, the iPhone *remediates* existing media, especially in relation to the genealogy of Apple products.[26] The iPhone merges the connectivity of the mobile phone with internet networking and a portable multimedia player by integrating the iPod. It becomes a platform for different media. Users can extend their mobile communicability to the online space and navigate digital information spaces on the move, and the social space of mobile phones can be linked with online social spaces. They can access mobile internet through the full web on smartphones—the tailored web or the mobile version of the web fitted for downsized screens and downloadable applications.[27] The multi-touch interface of the iPhone allows users to have easier and more intuitive operations, to instantly access applications and media functions, to scroll over content and to maximize or minimize screen pages with their fingers. One can customize an interface by selecting various applications to fit needs and tastes.

The iPhone, like all smartphones, makes it easy to call, SMS, MMS, and e-mail with simple gestures and to seamlessly interlink them with one another as well as with other applications. One can operate applications that link with a contact list and transmit content directly to people on the list. The iPhone also makes it easy to access the internet's social communication vehicles, such as SNS and instant messengers, as it incorporates the internet as an intra-medium—a medium within its multimedia platform. Function keys for phone calling and text messaging, once associated with earlier generation mobile media practices, are equally aligned with those of e-mailing, social networking and instant messaging, positioning themselves as multi-entry points for mobile social communication. Smartphone users can make phone calls and send text messages and also can freely access e-mail and SNS without logging in, or they can have a group chat in real time via mobile messengers where text messages, photographs and videos can be exchanged without charge. The iPhone, as part of the smartphone ecology, enables people to access online social spaces enabling multi-dimensional interactions while on the move.

The internet supposedly enables people to have a social space free from spatial and temporal constraints; people do not need to be co-located or to share the same time to use social communication tools such as e-mail, instant messaging and SNS. Internet connectivity allows person-to-person communication rather than place-to-place communication, leading to Barry Wellman's notion of "networked individualism,"[28] which implies that individuals can have less bounded and spatially dispersed social networks via computer-communications networks. When the tenuously knit, decentralized internet networks provide a new environment for social networking, an individual becomes a node for social connectivity. Social interactions and relationships take on individually based and fluid forms rather than place-based and sedentary ones. People can get various information and social resources that densely knit offline social networks can hardly offer. SNSs connect users who generate personal profiles online and thereby allow users to have various modes of communication oriented toward sociality and to build "ego-centric networks" on the web.[29] Since the smartphone serves as both a personal multimedia player and a connecting point to multiple social media on the internet, it further intensifies networked individualism such that an individual can develop or maintain his/her personal social networks without restraint.[30] This is amplified in the case of the iPhone and Apple's own streamlined version of social media through tailored social networking apps.

The smartphone, and especially the iPhone, mediate various communication media and render them accessible and manageable. Ralph Schroeder writes that mobile phones can be "regarded as part of a package of a larger communication and information diet of multimodal connectedness."[31] Smartphones as a convergent media can extend social media experiences by providing connecting points to various virtual communication spaces, and at the same time constitute a part of individual social communication environments in conjunction with existing media embedded in everyday life. Mobile social communications mediated by smartphones enable individuals to maintain states of *hyper-connection* and *hyper-awareness* of others. That is, users can engage in multiple social communication networks at any moment, continually access the various levels and scales of multi-layered communication contexts and expect that others in different contexts are always available in their virtual social spaces.

So in the context of this mobile media environment, how does the iPhone differ from other smartphones? Is there such a thing as an iPhone affect or practice? How does the iPhone shape social interaction and in turn how is iPhone use shaped by existing social relationships? What inter-media relations develop between existing forms and newly mediated forms? It is with these questions in mind that I explore the practices of iPhone users in Korea. Korea is a location with a relatively long history of smartphone innovation, and the iPhone represents both an extension of this phenomenon and also a marked departure. In particular, the adoption of the iPhone by users predominantly in their twenties and thirties in Korea means that the iPhone is also, in some ways, synonymous with this generation and their attendant media literacies and social media practices. Many comments can be found online from this generation expressing their enthusiasm for the iPhone. For instance, on Naver—one of Korea's major portal sites—there were more than 400 online communities related to the iPhone, and the number of the biggest community reached almost 500,000 members.[32] Considering young Korean iPhone users' experiences may give us an insight into the dynamic relationship between the iPhone, the changing mobile communication environments and everyday social interaction.

CASE STUDY—THE iPHONE IN KOREA

In order to gauge the social significance the iPhone as a personal communication system, I undertook in-depth interview surveys of iPhone users who used not only text messaging and calls but also instant messaging services and SNSs via their mobile devices. According to Digieco—Korea Telecom's research center—iPhone subscribers in their twenties accounted for 40 percent of total subscribers in October 2010.[33] They made up the most active age-group in adapting and using the iPhone for sociality. They also make up the age-group exposed to high-speed internet services and mobile phones since they were very young[34] and who show the highest usage rate of e-mailing, instant messaging and SNS use. *The 2010 Korean Survey on Internet Usage* reports that those in their twenties have a relatively higher usage rate of SNS service such as blogs (86.4 percent), online communities (84.8 percent), minihompy (85.9 percent) and microblogs (17.2 percent).[35] For this study, thirty-one iPhone users in their twenties living in the Seoul metropolitan area were recruited, and in-depth group interviews were conducted from April 2010 to February 2011. Of the thirty-one participants, fifteen were female and sixteen were male; twenty-three of them were students and eight were office workers or job seekers. Interviews were conducted for two to three hours in a semi-structured way that enabled participants to talk about their experiences openly. Interview participants talked about their motives for buying their iPhones, ways of using them, any perceived changes in their everyday media usage and their interaction and networking use of the iPhone. All interviews were transcribed and reviewed to collect the salient experiences shared by most interviewees, as well as to consider the experiential spectrum of iPhone usage. Pseudonyms are used throughout this chapter.

RECONSTRUCTING PERSONAL COMMUNICATION SYSTEMS

The iPhone tends to reflect and, at the same time, to cultivate the technological, economic and cultural needs of twenty-somethings in Korea. Before the launch of the iPhone, people in this age-group were accustomed to high-speed internet, and yet they could seldom access it with their mobile phones because few mobile devices allowing individuals to freely use the mobile internet were available, and charges for data transmission were too expensive. Of the thirty-one interview participants, twelve experienced the mobile internet and its applications via their iPod Touch devices and were waiting for an all-around iPod Touch equipped with the functions of mobile phones, digital cameras and GPS. Interview participants' motives for purchasing the iPhone and their usage of the device suggest that the iPhone met their desire for multi-functional mobile devices. Although their ways of associating multimedia functionalities with the iPhone slightly differ depending on personal tastes and media usage patterns, interview participants perceived the iPhone as a convergent medium that merges various media, including mobile phones, the iPod Touch, digital cameras, PMPs (portable multimedia players), GPS, notebooks, PDAs (personal digital assistants), books and so on.

Such multimedia functionality allows the iPhone to be used in multiple ways. On the one hand, the iPhone can function as a medium circumstantially tailored to meet personal needs. That is, the iPhone's media role can vary depending on an individual's situational needs. For instance, it can serve as a private medium in public places, as an MP3 player for listening to music while one prepares a document with a notebook or personal computer and as a portable computer when one needs urgently to view files attached to one's e-mail on the subway. It extends media experiences by supplementing other media in a parallel fashion or by substituting for them. On the other hand, as the iPhone integrates various media as its intra-media, it allows users to cross over to different media forms comfortably, to use them simultaneously or to interlink them in mutually complimentary ways. The iPhone, via apps, presents a particular version of the internet that differs from other smartphones. The iPhone's handy accessibility to and easy operability of multimedia functions as well as its stylish design make it an object of desire and also an icon with which to express one's style and identity. It can make iPhone non-users feel that they "lagged behind without it,"[36] and it "looks refined, trendy, and neat."[37] Here, we see that the symbolic functionality of the iPhone plays perhaps more of a role than its actual material functionality.

The iPhone as personal multimedia has reconfigured individual media experiences. With the new device, one can access existing media and yet use them in different ways. Although one might expect that users would employ the iPhone on a break or during "dead time" while on the move due to its mobility, in fact the iPhone serves as a medium for all times. Interview participants accessed the internet anytime and anywhere, whereas prior to having an iPhone they accessed it only at specific places with desktops wired to the internet. Consequently, total use of the internet has increased. In addition, the iPhone tends to replace a number of existing mobile devices. After the introduction of the iPhone with its multiple uses, existing mobile devices such as digital cameras,

MP3 players and notebooks were less likely to be carried and used. For instance, Jeeun, a 24-year-old female student, said, "I used to carry a notebook, but now I no longer take it with me. While I use the notebook when I stay at a place for long hours, I now handle most things such as e-mails with the iPhone." Personal notebooks transform into a relatively stationary medium in relation to the iPhone. Since the iPhone allows users to skip bothersome processes such as turning on and booting up computers to access the internet, one tends "to dispose of most matters with the iPhone"[38] and "to habitually fiddle about with it."[39] Interview participants rarely let their iPhones out of their hands and played with them continuously; they even took their iPhones with them to bed. They became dependent on both the iPhone's mobility and pervasive usability.

The iPhone's remediation of online social media transforms mobile social communication practices. Interview participants previously used mobile phones mostly for calls and SMS and to adjunctively use their camera, MP3 player and clock. With the iPhone, the device's main purpose changed to the use of SNSs, mobile messaging and e-mail. All of the participants did not consider phone calls to be the primary function of their iPhone. While they still use SMS, they often use other communication forms in online space. Although the main functions of existing mobile phones are neither discarded nor replaced, their significance has changed. For instance, interview participants mostly used mobile messaging, SMSs and SNSs, instead of calls; preferred free mobile messaging to pay-to-use SMSs; used mobile messaging more frequently than computer-based instant messaging; and checked out incoming e-mails more frequently than before. While the frequency of making phone calls tended to decrease, the amount of social communication increased with the constant use of instant messaging and SNSs. There have thus been significant modifications in the use of mobile communication media and social interactions. For instance, Rim, a 22-year-old female student, said that she doesn't "use phone calls unless there is an important matter or an emergency; I mostly use mobile messaging for talks."[40] Written messages are exchanged in near-synchronous and conversational mode and "e-mails become similar to dialogue,"[41] while "text messages become condensed and short, like online chatting messages."[42] Interview participants tended to use the social media that were compatible with the iPhone's intuitive interface. Social media such as Twitter and Facebook, which provide short, text-based systems or make it easy to check out and respond to messages almost in real time, are more favored than blogs and mini-homepages such as Cyworld that are perceived as heavy in text with little expectation of instantaneous feedback. Social media compatible with the device become more significant in mobile social communication; as they have various communication channels in their hands, respondents experienced a multi-layered mobile social communication space.

HYPER-SOCIALITY

Most interview participants used more than five applications to have conversations and share news, which included not only calls and SMSs, but also mobile messaging (e.g., *KakaoTalk, WhatsApp, Nateon*), SNS (e.g., Twitter, Facebook, Cyworld), blogs, internet phone calls (e.g., Skype, Viber), location aware SNS (e.g., *Foursquare*) and photo-sharing applications (e.g., *Instagram*). Individuals are able to use various social communication media; they can choose a specific medium suitable for interaction with certain communication partners or use various media to contact them. They can have social interactions in real time and extemporaneously have *optimal* communication channels appropriate for their purposes and situations.

While existing mobile phones allow users to have fluid interaction with remote people regardless of time, space and context,[43] iPhones allow them to move across diverse communication channels seamlessly. Henry Jenkins argues that in an era of digital convergence, users engage in "transmedia navigation" between media texts in multiple modalities and thus expand their range of media consumption activities.[44] The notion of transmedia indicates that there are variously dispersed "entry points" with which to access a media text, and that there are multi-modal versions of digital storytelling. The iPhone as a convergent medium enables users to experience another dimension of transmedia navigation. It provides multiple communication channels to access physically dispersed people and allows users to move across or "navigate" them. It helps to bring about *hyper-sociality*, or the state of being sociable in diverse virtual communication spaces. Such hyper-sociality has been practiced via iPhones in the following ways.

First, each channel can work as another's alternative. If one cannot get a response via mobile messaging, he/she moves on to text messaging. If one cannot get a response via an SMS, he/she uses calls and mobile messaging. If the mobile messaging system is shut off in the office, one can send messages via Twitter. Individuals can use a second-best medium if the best is not available.

Secondly, one channel can be used for the other's back channel. For instance, Sujin, a 23-year-old female student, moved back and forth between mobile messaging and Twitter when an unfamiliar person tried to get into her thread of conversation with friends.[45] Through this back channel of mobile messaging, she criticized that person's tactlessness, had a private talk and arranged a meeting.

Thirdly, transmedia uses via interlinked applications can occur spontaneously. One can take a picture with *Instagram*, a photo-sharing application, and simultaneously upload it to Twitter, Facebook, and *Foursquare*; he/she can continue to chat with friends while mobile messaging and social gaming. One can move across various media and have multi-dimensional interpersonal communication on the go. The iPhone deepens the state of the "tethered-self" that is, as Sherry Turkle writes, made possible by the way communication technologies are "always-on/always-on-you."[46] The iPhone provides entry points that allow users to maintain multiple-connectedness on the move and to engage in hyper-sociality where all sorts of mediated communication spaces are easily permeable to one another.

Users of the iPhone can have personally convenient communication spaces where they interact in real time or in a delayed way. They can initiate a conversation at any time, exchange words synchronously and asynchronously and respond to others instantly or later at a convenient time. For instance, Bong, a 29-year-old male office worker, said:

> In mobile messaging, I drop a word first. If there's a response, we'll talk, and if there is none, I would interpret it as either that they are busy or that they don't want to see me. While both I and my (communication) partner should simultaneously access mobile messaging for conversation, I would throw something on Twitter a little more leisurely, and if someone takes the bait, I would play with them.[47]

Depending on others' feedback, he can have either near-synchronous or asynchronous communication, and his words can take the form of either immediate or deferred dialogue that waits for a response. He flexibly takes on different modes of communication and heightens the possibility of social interaction. Since one extends the resources of shared experience through continuous "word dropping," providing updates of his/her state of being, and checking incoming messages in real time via the various communication channels of the iPhone, social interaction can occur at any time. There are always people to talk to and play with within such a mobile social communication space. Mobile multi-connectedness extends accessibility to others and reinforces ego-centric communication practices.

One can also shift between multiple channels and find communication partners fit for their mood, situation and communication purpose at that moment. Ri, a 21-year-old female student, stated:

> I feel comfortable to talk with people I meet on Twitter or the other online sites. While it is a bit difficult for me to talk about the exam with people who took it, it is really comfortable for me to talk about it with someone I share no stake with. If I express via Twitter my feeling at the moment, for instance, "it is hard to take what someone says to me at my part-time job," I get consolation from people who have similar experiences. I can get consolation, converse about really trivial things, share anger about news or deliver news with its link. On the other hand, I would rather use SMS and *KakaoTalk* for instrumental purposes because I am taking this semester off from school. While in school, I always had something to talk about with my schoolmates since I met them every day and conversed with them via SMSs and *KakaoTalk*.[48]

Ri spoke of her personal feelings and thoughts to networked friends with whom she had little chance to directly encounter in everyday life on an open SNS, and received immediate responses and emotional support from them. She felt comfort in being anonymous and enjoyed small talk. Yet, she also used one-to-one text messaging or mobile messaging to exchange messages with offline acquaintances for instrumental purposes. She accessed each communication channel for different purposes and with different expectations. As one fluidly moves across multi-communication channels, he/she can build a personal feel for each channel and have an expectation for certain levels and modes of communication in it. Thereby, one can acquire an ability to choose the most appropriate communication channel and partners in accordance with her/his needs and conditions of the moment. The iPhone provides entry points to diverse social networking and at the same time works as a technology-managing hyper-social device. Accordingly, new practices of social interaction and networking emerge.

THE TECHNOLOGY OF MANAGING SOCIALITY

Through the mobile connectivity of the iPhone, one can access not only direct friends but also friends of online friends. One can connect to pre-existing friends as well as people with whom one has little chance to interact socially and yet is linked virtually. Social media on the web are used not only to maintain existing relationships but also to interact with people in larger sets of weak ties and latent ties. Carolyn Haythornthwaite argues that the introduction of a new social networking medium can create latent ties "that are technically possible but not yet activated socially." While this new medium can be actively used by those more strongly tied, it can also lay an infrastructure for latent ties between formerly unconnected others, which can turn into weak or strong ties by social interaction.[49] Nicole Ellison et al. also emphasize that SNSs may be used to maintain pre-existing relationships and yet provide "new pathways of communication" with unknown others.[50]

The iPhone as a convergent medium allows users to stay connected to the networks of strong ties as well as to the online networks of weak ties and latent ties and to extend the personal circle of virtually co-present others. While the mobile phone used to contribute to maintaining strong ties with family members and close friends, the iPhone makes anyone on the contact list a possible conversant via mobile messaging and conveys in real time the updates of registered friends on online social media. Most interview participants said that they had conversations not only with their close friends, classmates and fellow workers, but also with others who shared common interests and with those whom they hadn't known previously. Ho, a 27-year-old male student, said, "I have social ties with strangers via SNS and share a close bond with those people whom I know but have rarely contacted."[51] Taek, a 21-year-old male student, had a chance to meet friends from junior high school who had drifted apart,[52] and Mi, a 20-year-old female student, said, "I contact friends with whom I used to be close to but lost track of" through her iPhone.[53]

The iPhone provides another interlinked communication channel to contact distant people and enables users to get a sense of the connected presence of technologically mediated friends. But unlike other smartphones, the affordances of iPhone contact are situated through, and by, the customization of apps especially SNSs like minihompy. In sharing phatic communication and small talk via the iPhone and features such as the SMS "continuum" in which various conservations are displayed through SMS bubble texts, different types of mobile etiquette are performed. Like other smartphones and mobile connectivity in general, the iPhone enabled a heightened awareness over multiple communication networks, such that users become more talkative (virtually) and have more social interaction. Yet most interview participants said they became more socially active after using the iPhone.

As interview participants gained longer lists of closely connected friends and maintained larger networks, they learned how to manage such expanded networks. Interview participants began to acquire techniques for managing the multiple social networks mediated by their iPhones. Such management techniques include *differentiating, customizing, adjusting* and *numbing*. Here we see how expectations of use played an important role in how and whom users interacted with and when. For instance, Hye, a 25-year-old female student, "uses *WhatsApp* for people I met on Twitter and became intimate offline friends with, *KakaoTalk* for my old friends who use smartphones, Facebook and Twitter for those people I become intimately acquainted with online, calls and SMSs for offline friends."[54] When people use multiple communication channels, across platforms and through iPhone apps, they tend to *differentiate* among their interactants based on the characteristics of social ties as well as the main communication channel for contact. *Interactants* are classified into many categories, such as people with strong ties and people with weak ties, people in formal relationships and those in private relationships, offline friends and online friends, smartphone users and non-smartphone users and so on. These categories are further subdivided depending on the most used communication channel for interaction, such as those friends they usually interact with by mobile messaging, those they converse with via SNS, those they exchange emails with and so on. iPhone usage thus facilitates a detailed breakdown of their social relationships.

iPhone users can also *customize* their social interactions. Considering the characteristics of each communication channel, that is, its accessibility, temporality, modality and main interactants, one can choose the right channel to interact with her/his communication partner and control the level and scale of social interaction. For instance, Eun, a 24-year-old male student, preferred text-based communication channels due to his heavy accent.[55] Chul, a 29-year-old male graduate student, used free mobile messaging to talk about many things with intimate friends and SMSs to contact friends who were not close.[56]

Stefana Broadbent and Valerie Bauwens argue that, "multiple channels are being used with the same partners for different situations and contents".[57] While people tend to use diverse channels for maintaining strong ties, they use a much smaller range of communication resources for weak ties. Similarly, Ellison et al. suggest that people are probably more likely to make use of multiple methods of interaction to maintain their relationships with close friends than those with weak ties.[58] Rather than reconfirming such previous studies, interview participants showed another tendency.

Although close family members and friends could be contacted via calls, SMS and mobile messaging, they were not necessarily registered as Twitter friends. One could meet Twitter friends in various interlinked applications such as Facebook, Foursquare and so on, and even develop an online relationship into an intimate offline relationship. Interview participants were more likely to find a proper communication channel or a *niche* where they could interact most comfortably and efficiently, depending on their communication partners as well as their surrounding conditions. As the connectivity of the iPhone allows various circles of social networks to vie for one's attention and facilitates the blurring between online and offline relationships, the forms and density of mediated social relationships become more complicated, and users develop ways to manage such complicated social relationships. One can assign the most suitable communication channels to varied social relationships and make the most of each channel to meet personal needs for communication.

As one can access diverse communication networks via the iPhone, one can also survey personal communication networks and become aware of communication flow among these networks. One can customize the structure of personal connectivity and *adjust* accessibility on personal networks as well as on privacy levels. For instance, Taek, a 21-year-old male student, said:

> Since my family member subscribed to Facebook, I made another account. I do not like to mix offline people with my online friends. I do not have actual friends on Twitter, but on Facebook, I let information flow out of Facebook but keep Twitter's message from flowing into Facebook. My friends do not know about my Twitter account. I daily write funny things on Twitter. It would be embarrassing to show them to my friends.[59]

Many interview participants did not like mixing their social networks. Despite the expandability and interconnectivity of mediated social networks, they tried to compartmentalize different social networks, to keep certain social circles from flowing into another and to minimize embarrassment from mixed audiences. Like Taek, some interview participants kept their offline acquaintances from accessing their Twitter or Facebook accounts, managed different networks by grouping them or made second accounts for private interactions. Other interview participants started to review their contact lists after using the iPhone because this list was interlinked with mobile messaging and they didn't want strangers to pop up in their mobile messaging. When managing diverse communication networks, people faced the question of how far they would open their personal relationships and private musings, as well as how they could manage various social networks without confusion and gain control over personal communication flows.

Due to the iPhone's interface and platforms, users have to manage more communication channels, messages and different methods of interaction (such as the haptic interface) than before. Interview participants made efforts to adjust reaction times to incoming messages in real time and tried to routinize their message checkups and to regulate the interaction. By using iPhones, they spent more time socializing and had to deal with an influx of messages. They also developed a sense of *numbing*; they became insensitive to incoming messages and felt less obligation to respond immediately. While the technological affordances of the iPhone allow multiple real-time interactions, people become more accustomed to no response or delayed response.

For instance, Won, a 20-year-old male student, said, "The alarms of incoming text messages are counted not much differently from those of Twitter and Facebook messages."[60] Since so many incoming messages piled up on his iPhone, he often missed new text messages and didn't reply. The iPhone, with its mobile hyper-media interface, enables users to check and respond to various forms of messages in near real time and yet, ironically, facilitates numbing to incoming messages. Users are sometimes overwhelmed by incessant incoming messages and begin to justify their delayed responses. Although they cannot totally ignore those messages in order to maintain their relationships, they learn countermeasures through experience. They look forward to immediate responses, and yet they try to remain aloof. While the iPhone's portability and instant connectivity, as a customized form of smartphone, enables users to have interactions anytime and anywhere and to heighten connected presence, hyper-connectedness also causes distractions and insensitivity to incoming messages.

CONCLUDING REMARKS

This chapter has examined how the iPhone shapes personal communication systems and people's communication practices and social interactions in the Korean context. With an already abundant and innovative technoculture, the entry of the iPhone has represented a particular shift in the smartphone phenomenon in Korea. The specific affordances of the iPhone—both perceived and actual—have become a conspicuous part of twenty-somethings' lives in Korea. As Korean people use iPhones in their everyday life, their mobile social interactions become multi-layered. The forms of existing mobile communication, such as calls and SMSs, obtain different social significance, confirming Walter Ong's point that "the advent of newer media alters the meaning and relevance of the older."[61] The iPhone phenomenon in Korea has facilitated and accompanied the rise of mediated social interactions and consequent concerns about how to manage multiple communication channels including social networks. This hyper-connectivity is highlighted in my case study of Korean iPhone users. Without the iPhone, interviewees would feel "nervous, bored and empty"[62] and/or "worried friends would be anxious."[63] The iPhone's absence is experienced as a source of great anxiety, engendering a loss of presence. This feeling was exemplified in one interviewee's comment that the iPhone "cannot be separated from my hand."[64]

While part of a broader smartphone ecology, the iPhone proffers a particular version of everyday media practice in Korea whereby different communication media are incorporated and internalized within its interface, which in turn enhances already prevalent hyper-sociability. Participants used the iPhone to fluidly navigate interlinked communication channels, engage in social interactions in various communication sites, orchestrate multiple communication channels, manage expanded social networks and find niche communication channels suitable to their current context. Smartphone-oriented sociability closely associates social relationships with the use of communication channels, and it differentiates them. Smartphone-oriented sociability also tends to isolate non-iPhone users and to increase the perception that one should handle it well and efficiently. As people have become more dependent on the technological capabilities of the iPhone, users manage complex layers of social interaction and learn to deal with scalable connectivity and expanded social networks.

NOTES

1. *2010 Survey on Wireless Internet Usage* (Seoul: Korea Internet & Security Agency, 2010).
2. "Gooknae iPhone yiyongja 50man dolpa (The number of iPhone users exceeded 500,000)," *Money Today* , 1 April 2010, http://www.mt.co.kr/view/mtview.php?type=1&no=2010040111470278777&outlink=1 (accessed 2 April 2010).
3. Korea Communications Commission, "Smartphone gaipja 1000man dolpa, smart sidae bongyeok gaemak (10 million smartphone users [and] the beginning of the genuine smart era)," 24 March 2011.
4. *2008 Internet Report Volume 4* (Seoul: ETRC, 2008).
5. See Daewon Kim, *Apple Shock* (Seoul: The Nanbiz, 2010).

6. *Maeil Business Newspaper,* http://news.mk.co.kr/newsRead.php?year=2010& no=647351 (accessed 25 November 2010).
7. *The Seoul Shinmun,* http://www.seoul.co.kr/news/newsView.php?id= 20101126018009 (accessed 26 November 2010).
8. *Etnews,* http://www.etnews.com/news/detail.html?id=201011250089 (accessed 28 November 2010).
9. Daewon Kim, op cit., 206.
10. Jungnam Kim, *The Secret of Apple's Success Story* (Seoul: Hwangkeum Bueongi, 2010).
11. Yongseok Choi, *Strategies of the iPhone and the iPad* (Seoul: Arachne, 2010).
12. Yeonghan Kim, *Steve Jobs' Creative Charisma* (Seoul: Readers Book, 2010).
13. Inseong Kim, *The Decline of the Korean IT Industry* (Seoul: Bookhouse, 2011).
14. Heonyong Park, *The Unsmart Smart Wars* (Seoul: Donga E&D, 2010).
15. Jeffrey Boase, "Personal networks and personal communication," *Information, Communication & Society* 11(4), 2008: 490–508.
16. *iPhone Doip Ilnyun (1 year after the launch of iPhone)* (Seoul: Digieco, 2010), 6.
17. "1000 mangoji neomeun kakaotalk 'kookmineui app' wodduk (The number of Kakaotalk subscribers exceeded 10 million and became a 'people's app')," *Money Today,* 3 April 2011, http://www.mt.co.kr/view/mtview.php?type=1 &no=2011040311254518416&outlink=1 (accessed 4 April 2011).
18. James E. Katz and Mark A. Aakhus, eds. *Perpetual Contact: Mobile Communication, Private Talk, Public Performance* (Cambridge, UK: Cambridge University Press, 2002).
19. Mizuko Ito and Daisuke Okabe, "Technosocial situations: emergent structuring of mobile e-mail use," in *Personal, Portable, Pedestrian: Mobile Phones in Japanese Life* (Cambridge, MA: MIT Press, 2005), 264.
20. Christian Licoppe, "'Connected' presence: the emergence of a new repertoire for managing social relationships in a changing communication technoscape," *Environment and Planning D: Society and Space* 22, 2004: 135–156.
21. Ruth Rettie, "Mobile phone communication: extending Goffman to mediated interaction," *Sociology* 43(3), 2009: 421–438.
22. Ibid., 434.
23. Hans Geser, *Towards a Sociological Theory of the Mobile Phone,* http:// socio.ch/mobile/t_geser1.htm (accessed 14 April 2008); Richard Ling, *New Tech, New Ties: How Mobile Communication Is Reshaping Social Cohesion* (Cambridge, MA: MIT Press, 2008).
24. Ichiyo Habuchi, "Accelerating reflexivity," in *Personal, Portable, Pedestrian: Mobile Phones in Japanese Life* (Cambridge, MA: MIT Press, 2005), 167.
25. Kenneth Gergen, "The challenge of absent presence," in *Perpetual Contact: Mobile Communication, Private Talk, Public Performance* (Cambridge: Cambridge University Press, 2002), 240.
26. David Bolter and Richard Grusin, *Remediation: Understanding New Media* (Cambridge, MA: MIT Press, 1999).
27. Anne Kaikkonen, "Mobile internet: past, present, and the future," *International Journal of Mobile Human Computer Interaction* 1(3), 2009: 29–45.
28. Barry Wellman, "Physical place and cyber-place: the rise of networked individualism," *International Journal for Urban and Regional Research* 25, 2001: 227–252.
29. danah boyd and Nicole Ellison, "Social network sites: definition, history, and scholarship," *Journal of Computer-Mediated Communication* 13(1), 2007, http://jcmc.indiana.edu/vol13/issue1/boyd.ellison.html (accessed 20 January 2009).

30. Manuel Castells, Mireia Fernandez-Ardevol, Jack Linchuan Qiu, and Araba Sey, *Mobile Communication and Society: A Global Perspective* (Cambridge, MA: MIT Press, 2007); Scott W. Campbell and Yong Jin Park, "Social implications of mobile telephony: the rise of personal communication society," *Sociology Compass* 2(2), 2008: 371–387.

31. Ralph Schroeder, "Mobile phones and the inexorable advance of multimodal connectedness," *New Media & Society* 12(1), 2010: 77.

32. "Digital gigido fanclub sidae (The era of digital device fandom)," *Yeonhap News*, 22 February 2010,http://news.naver.com/main/read.nhn?mode=LSD &mid=sec&sid1=101&oid=001&aid=0003133230 (accessed 23 February 2010).

33. *iPhone Doip Ilnyun (1 year after the launch of iPhone)* (Seoul: Digieco, 2010).

34. According to the Korea Communications Commission, the rate of households with internet access reached over 70 percent in 2002. In the same year, the mobile phone subscription rate for the Korean population was 63 percent.

35. *Survey on Internet Usage* (Seoul: Korea Internet & Security Agency, 2010).

36. Young, a 26-year-old female job seeker, interview by author, Seoul, 11 May 2010.

37. Ho, a 27-year-old male student, interview by author, Incheon, 18 January 2011.

38. Jin, a 26-year-old male student, interview by author, Seoul, 11 May 2010.

39. Chul, a 29-year-old male graduate student, interview by author, Seoul, 11 January 2011.

40. Rim, interview by author, Seoul, 8 February 2011.

41. Roo, a 23-year-old female student, interview by author, Seoul, 9 February 2011.

42. Younga, a 23-year-old female student, interview by author, Seoul, 14 February 2011.

43. Masao Kakihara, Sorensen Carsten, and Wiberg Mikael, "Fluid interaction in mobile work practices," First Global Mobile Roundtable (Institute of Innovation Research, Tokyo, Japan), 29–30 May 2002.

44. Henry Jenkins, *Convergence Culture: Where Old and New Media Collide* (New York: New York University Press, 2006), 215.

45. Sujin, interview by author, Seoul, 9 February 2011.

46. Sherry Turkle, "Always-on/always-on-you: the tethered self," in *Handbook of Mobile Communication Studies* (Cambridge, MA: MIT Press, 2008), 121–137.

47. Bong, a 29-year-old male office worker, interview by author, Seoul, 3 May 2010.

48. Ri, a 21-year-old female student, interview by author, Seoul, 9 February 2011.

49. Caroline Haythornthwaite, "Social networks and internet connectivity effects," *Information, Communication & Society* 8(2), 2005, 137.

50. Nicole Ellison, Cliff Lampe, Charles Steinfield, and Jessica Vitak, "With a little help from my friends: how social network sites affect social capital processes," in *A Networked Self: Identity, Community and Culture on Social Network Sites* (New York: Routledge, 2011), 130.

51. Ho, interview by author, Incheon, 18 January 2011.

52. Taek, a 21-year-old male student, interview by author, Seoul, 11 February 2011.

53. Mi, a 20-year-old female student, interview by author, Incheon, 26 January 2011.

54. Hye, a 25-year-old female student, interview by author, Seoul, 12 February 2011.

55. Eun, interview by author, Incheon, 26 January 2011.
56. Chul, interview by author, Seoul, 11 January 2011.
57. Stefana Broadbent and Valerie Bauwens, "Understanding convergence," *Interactions* 15(1), 2008: 26.
58. Ellison et al., op cit.
59. Taek, interview by author, Seoul, 11 February 2011.
60. Won, interview by author, Seoul, 28 April 2010.
61. Walter Ong, "Oral residue in Tudor prose style," in *An Ong reader: Challenges for Further Inquiry* (Cresskill, NJ: Hampton Press, 2002), 314.
62. Gi, a 26-year-old male student, interview by author, Incheon, 18 January 2011.
63. Yeon, a 26-year-old male student, interview by author, Incheon, 10 January 2011.
64. Jung, a 23-year-old female student, interview by author, Seoul, 28 April 2010

Part II

iPhone as a Platform and Phenomenon

6 iPhone Photography
Mediating Visions of Social Space

Daniel Palmer

INTRODUCTION

Cameras have colonized the mobile phone over the past decade. Today—fifteen years after Swiss inventor Philippe Kahn jerry-rigged a mobile phone with a digital camera to circulate images of his newborn daughter—no phone is complete without a camera, and more phone cameras are now sold than any other kind of camera. Indeed, the mobile phone company who can now lay claim to being the world's largest camera maker, Nokia, has reportedly put more cameras into people's hands than in the whole previous history of photography.[1] Most citizens of the developed world now carry at least one camera at all times, capable not only of recording and displaying images but also instantly sharing them, via the internet or messaging services. This radically new development has generated equal measures of enthusiasm and anxiety. For instance, as much as phone cameras enable forms of intimacy at a distance and are celebrated as devices of on-the-spot "citizen" photojournalism—such as during the 2009 Iran elections, or following the 2011 Japanese earthquake—they are also intertwined with larger social anxieties about privacy in an age of digital dissemination. This chapter looks at the iPhone as the exemplary instance of the mobile phone's convergence with the camera, as a device that casts the potential capabilities and affordances of camera phones in stark relief. It considers the distinctive qualities of iPhone photography and explores both everyday and innovative uses of the device by ordinary users as well as artists and activists.

THE iPHONE PHOTOGRAPHY EXPERIENCE

In significant respects, the iPhone continues a number of trends established by earlier camera phones. The iPhone arrived into an ecology of images and image-making in which snapshot photography has shifted away from a print-oriented mode to a transmission-oriented, screen-based mode.[2] The nature of digital snapshots has already attracted considerable attention, and there is widespread agreement among researchers that such images are both more intimate and mundane than earlier forms of personal photography.[3] A variety of studies have also shown that camera phones are associated with ephemeral and spontaneous photographic practices. Since the camera phone is always available, it supports a particularly mobile and informal way of taking and consuming images, including visual jokes and functional visual notes (such as an image of a potential gift to discuss with a partner or a snapshot of a restaurant meal in lieu of a verbal description). Daniel Rubenstein has described the phenomenon as "visible speech."[4] Likewise, Mikko Villi has suggested that photographs exchanged by MMS and instant messaging function to achieve a form of communicative presence that extends the mediated presence offered by traditional photography. Villi further argues that the camera phone enables a form of communication that resembles a pictorial conversation or "visual chit chat."[5]

As part of direct communication, between friends and lovers, camera phone photography is often intimate—even erotic in the controversial case of "sexting." Anna Reading and others have argued that camera phones articulate complex gendered identities that traverse established boundaries of men's and women's experiences.[6] With the iPhone 4, the addition of a front as well as a rear-facing camera means the screen can alternate between a window and a mirror. Specifically designed for Apple's FaceTime and video communication like Skype, the front-facing camera also means that self-portraits can be taken more easily. Holidaying individuals and couples can often be seen using this function in picturesque locations, overcoming the need to imagine a frame of vision. Thus, at the same time as privileging the everyday and the transient, the camera phone is also a "life recorder," generating a fragmented archive of a personal trajectory or viewpoint on the world. This has become increasingly important with the intersection of the camera phone and photo-sharing sites online.

The iPhone is neither the first nor the highest quality camera phone. Indeed, the iPhone's camera unit has notoriously lagged behind other advanced phone cameras in the megapixel race. The original iPhone launched with a fixed-focus lens and a paltry 2-megapixel sensor, small even by 2007 standards.[7] And yet the iPhone is now the most popular type of camera on Flickr, responsible for more of the uploaded pictures than any other single type of camera by the likes of Canon or Nikon. This is perhaps all the more remarkable considering that when Apple introduced the iPhone, it promoted the device as a combination of "a revolutionary mobile phone," "a widescreen iPod" and "a breakthrough internet device." The camera, it seemed, was included almost by default (although the "photo management application" was singled out for note). The camera was neither a novelty nor a key feature. Indeed, it is just one among many of the iPhone's embedded *sensors*, such as the accelerometer, digital compass, gyroscope, GPS and microphone. Yet as we will see, the transformation of the camera into a digital sensor has profound implications for how we might understand photography.

What makes the iPhone camera so distinctive? Why is it that scores of websites and even books have been published on what has been dubbed iPhoneography?[8] The first answer lies in the high-resolution, color touch-screen of the iPhone. Mobile phones became recording and consuming devices of media in the early 1990s, but in the process of seamlessly melding mobile, computing and internet cultures, the iPhone turned the phone into an attractive display device. Effectively, the iPhone imported the values and technologies that Apple has always been associated with in relation to personal computing: a graphical interface and strong design values that pay attention to ease of use and aesthetics. While much has been written about the touchscreen interface, the quality of the large color display has been equally important to the iPhone's success compared to other smart-phones like the BlackBerry. The improved high-resolution screen on the iPhone 4 even inspired Apple to build High Dynamic Range (HDR) into the camera software, in which scenes are automatically exposure-bracketed. Apple computers have traditionally been favored by visual professionals—designers, artists, photographers and so on—and anecdotal evidence suggests that the iPhone has also been adopted disproportionately by people working in such industries.

The experience of using the iPhone's native "camera app" is revealing. When the built-in app is activated, a simulated aperture blade opens on the screen to show the scene viewed by the tiny fixed aperture lens on the rear of the phone (or, since the iPhone 4, on the front). As it happens, the iPhone uses a CMOS sensor, which more or less "wipes" the shutter across the sensor like a scanner rather than the circular aperture of a traditional camera. Likewise, the "button" at the bottom of the screen with which you take a photograph—or commence a video recording—appears as a camera icon that resembles a traditional single lens reflex (SLR) film camera. When pressed, a photograph is taken, and an artificial shutter click is heard. Instantly, in a brief animation, the captured image quickly shrinks and jumps into a little window at the lower left of the screen, and the camera is ready to take another picture. To view the pictures, by pressing the lower left image or opening the native "photo app," we find the "camera roll." From here we can move through the images, trash them, or choose from a series of listed options: "Email photo, MMS, Assign to Contact, Use as Wallpaper, Tweet, Print." There is a strong sense in all of this that the world is readily available for visual consumption. The camera has always been a device to collect visual impressions of the world, but the iPhone takes this to another level. Scenes are quickly grabbed—the fast processor gives the iPhone less shutter lag than most camera phones—and just as quickly and efficiently deleted, shared and archived.

The iPhone is carefully designed to make viewing photographs an emotionally satisfying experience. Aside from the quality of the screen resolution, the touchscreen enables swiping and pinching to zoom in and out and around the image. The native photo app on the iPhone is a scaled-down version of the very popular consumer software application iPhoto, and synchronizes "albums" and "events" (and more recently "faces" and "places"). It turns the iPhone into a wallet of archived photographs that can be passed around at the dinner table with friends. The orientation of the image is automatically adjusted. That is, someone viewing photos on their iPhone can rotate the device 90 degrees, from portrait to landscape, and the sensing device will detect the movement and change accordingly. These acts have quickly become commonplace—and widely imitated by a variety of other phone manufacturers—yet their importance should not be diminished. What it enables is a form of embodied visual intimacy. Touch has long been an important, but neglected, dimension in the history of photography.[9] The acts of holding and passing around physical prints, such as running one's hands over images of absent loved ones, or carrying torn paper images in a wallet, has long been part of the practice of viewing personal photographs. The iPhone, held in the palm of the hand, reintroduces a visual intimacy to screen culture that is missing from the larger monitor screen. And since the release of the iOS4 in 2010, the touchscreen also enables touch-focus and zoom capacity in the act of taking photographs.

The second reason that iPhone photography has become a phenomenon is due to the proliferation of photo "apps" available for download through the Apple iTunes App Store (see Chris Chesher in this collection). The native iPhone camera app is just one piece of software among hundreds that can engage the camera hardware on the phone. One of the much-remarked qualities of the iPhone is the innovation behind the developments of free or cheaply downloadable apps, and photography applications are no exception. Photo apps have two main purposes: first, to make images more "artistic" or aesthetically appealing, and second, to facilitate the distribution or publication of photographs via the internet. Many do both. In the first case, the most popular iPhone photo apps mimic the lo-fi look of cheap film-based cameras such as Lomo or Polaroid.[10] The development of such apps, like the more general embrace of the plastic toy camera, is both a nostalgic reaction against the monotonously—at least until recently—slick perfection of images produced by digital cameras, and arguably a way to obscure the relatively poor quality of the iPhone's camera. What they entail is a form of "processing" of the JPEG image in a manner that emulates a retro look, typically including a square format, faded tones, vignetting and chromatic aberration effects. While often gimmicky, the effects can be surprisingly effective. One of the most popular of such apps is *Hipstamatic*, whose very name suggests its target market among a young, hipster crowd. When *Hipstamatic* loads, the iPhone's screen is transformed into a cheap analogue camera with a small viewfinder. The turnover icon at the bottom flips the camera so you can see the front, allowing you to scroll through different lenses (three are included), choose a film and (simulated) flash type. *Hipstamatic*'s variables are designed to simulate the looks from certain films and generally give a 1970s feel to the images—right down to the simulated paper border. At a cost of US$1.99, it consistently ranks among the most downloaded iPhone apps. Under the "Share" option on *Hipstamatic*, in addition to e-mailing, users have the options to "Share on Facebook, Tweet on Twitter, Post to Tumbler or Upload to Flickr." In other words, such apps are synchronized to dominant photo-sharing websites.

Needless to say, popular blogging platforms like Tumblr and WordPress as well as Twitter have their own iPhone apps that allow for quick uploads of photographs directly from the iPhone. As Gerard Goggin has observed, "Theorists and promoters of citizen journalism offer the cell phone camera user as the epitome of this new paradigm."[11] Indeed, the camera phone was essential to the reporting of such events as the 2005 London Bombings, the 2009 Iran elections and the 2011 Japanese Tsunami. In a major shift, mass media news organizations now routinely present images and video footage of scenes before they have verified the accuracy of the content, with a disclaimer along the lines of "purporting to show." Many media organizations have released apps to facilitate this integration of user-generated content into their programming.[12] The best camera is the one you have on you, as the saying goes—and when it comes to breaking news, that may be the iPhone (although the chances of this are less likely outside the West, due to their relatively high cost).[13] By contrast, in 2010 professional photojournalist Damon Winter opted to use an iPhone and the *Hipstamatic* app as a matter of aesthetic choice to take a series of photographs in Iraq for the *New York Times*. Defending his award-winning series against accusations of manipulation, Winter said he usually used the iPhone to take pictures of his cat but felt it also enabled him to tell a personal story about US soldiers in a new light:

> The beauty of a new tool is that it allows you to see and approach your subjects differently. Using this phone brought me into little details that I would have missed otherwise. The image of the men resting together on a rusted bed frame could never have been made with my regular camera. They would have scattered the moment I raised my 5D with a big 24–70 lens attached. But with the phone, the men were very comfortable. They always laughed when they saw me shooting with it while professional cameras hung from my shoulders.[14]

Strangely, while embracing the intimacy of the snapshot aesthetic, Winter makes no reference to the retro palette of the pictures, which evoke the Vietnam War and the heroic age of photojournalism.

A third feature of the iPhone that makes its photography distinctive is a consequence of the phone's GPS capability. The iPhone automatically tags photographs with their location, allowing images to be browsed and arranged geographically. In the "Places" feature on the photo app (or on iPhoto), a Google map shows where all the photographs were taken. A number of photo-sharing websites, such as Flickr (www.flickr.com) and Panoramio (www.panoramio.com), now provide large collections of publicly available, accurately geo-tagged images. Indeed, uploaded iPhone images may even be repurposed for tasks involving navigation, as a kind of byproduct. The significance of GPS for photography is only just being explored, as we will see below. Once again, the iPhone is popularizing a development that is likely to become standard in digital cameras, with important social consequences.

Finally, and most radically, the iPhone signals a shift in thinking about photographs as being primarily about *representation* to thinking about photographs as *information*. As noted earlier, the camera is just one among many of the iPhone's embedded sensors. In a sense, the camera is just one of many "data collection features," and while this opens up possibilities for "participatory sensing" (see below), it also reminds us that digital images are increasingly aggregated and analyzed by computers rather than just humans. This is overtly the case in so-called QR (Quick Response) codes— square symbols found on advertisements or alone in the urban landscape that can be scanned by the mobile camera (and automatically generated by Google). The barcodes typically contain store addresses and URLs that are usually translated into promotions or other commercial messages. Downloadable reader apps like *QuickMark* and *QR App* turn the city into an interactive experience that for the most part so far appear to extend the logic of commodifying public space.

THE SOCIAL CONTEXT OF CAMERA PHONES: URBAN SPACES AND SURVEILLANCE

Mobile and online media have entrenched ideas about the need to control images for fear of voyeuristic practices like "upskirting," not to mention pedophilia and terrorism. In the Australian media, hardly a week goes by without the latest social media scandal involving the inappropriate distribution of personal imagery, and the terms "digital image" and "loss of control" have become almost synonymous. This has inevitably bred a certain mistrust of professional photographers working in public. Even Rex Dupain, the well-known son of Australia's best-loved modernist photographer, Max Dupain, has recently complained of being treated like a predator or criminal when photographing at the beach.[15] This is despite the celebrated history of photographers who have made a virtue of working without the explicit permission of their subjects, often working in exposed places like the street to explore the paradox of intimacy in public.[16] Ironically, photographers are under siege at a time when voyeurism has been turned into an entertainment form, and when voluntary self-surveillance has become a leisure pastime. On sites like Facebook, people actively dissolve conventional public-private boundaries, posting intimate details and pictures of themselves as a form of performative self-representation.[17] Attacks on photographers are surely symptomatic of deeper cultural anxieties that manifest as "bodily privacy" but have the unfortunate consequence of naturalizing the monopoly of official surveillance by the corporatized state. In this sense they rupture what Ariella Azoulay has dubbed the "civil contract of photography," the possibility for non-state civic interaction allowed by the invention of the camera, and the capacity of photography to circulate political claims.[18]

In key respects, debates about phone cameras parallel the public paranoia that accompanied the rise of portable cameras in the late nineteenth century. The 1898 the *British Journal of Photography* moaned of "the hand-camera fiends who 'snap-shot' ladies," and for a period Victorian photographers required permits to take pictures of people in London's parks.[19] New technologies invariably generate anxiety. However, the assemblage of phone cameras and social media networks is particularly potent, and the introduction of GPS technology complicates things even further, generating new ideas such as "locational privacy."[20] With important exceptions, people living in Western democracies have traditionally enjoyed the right to be able to move around anonymously. Like face recognition software installed in security cameras, GPS-enabled devices threaten this freedom, exposing users to potentially new forms of surveillance. Hence, the recent controversy surrounding iPhone's hidden "tracking" file—a database of the user's location stretching back for months—and legitimate concerns that Apple was secretly yielding data for location-targeted advertising (see Kate Crawford in this collection).

CREATIVE USES OF iPHONE PHOTOGRAPHY

Given its cultural impact, it is hardly surprising that a number of artists have become intrigued with the iPhone. While few have made it their primary focus, several high-profile artists have produced bodies of work that explore the aesthetic possibilities related to the immediacy and low-resolution rawness of the images. Rob Pruitt's *iPruitt* (2008) was among the first, "a stream-of-consciousness photo diary" of over 2000 prints hung in grid fashion—of largely banal subjects that prominently featured moments of intimacy and moments of voyeurism (tellingly, the poster for the exhibition featured a stranger's blue jean crotch captured on an airport shuttle bus).[21] Pruitt, better known as an installation artist inspired by Andy Warhol's pop legacy, supplemented an elegiac feel by introducing leaves on the gallery floor and plastering images on the wall outside of gallery. Rather than single images, the inkjet prints were sold in chronological chunks; for instance, one collector apparently bought everything from September.[22] In a more documentary mode, Joel Sternfeld—best known for luscious 8- by 10-inch view camera work such as the influential *American Prospects*

(1987)—used the iPhone to produce a book project titled *iDubai* (2010). While the images further Sternfeld's longstanding wry critique of consumerist "utopias," now figured in the malls of Dubai, the project draws attention to the iPhone as novelty device. For Sternfeld, the Baudelairean figure of the *flâneur* is replaced by "a wired wanderer" who snaps digital images on the fly.[23] Reflecting on the project, Sternfeld speaks of a new grammar of photography, as it shifts from a "privileged discrete act to something more continuous and generic."[24] A common refrain that accounts for the appeal of the iPhone to artists is that it is unobtrusive, almost invisible, compared to larger cameras.

An important context for experiments with camera phones are those artists whose work questions simplistic dynamics of surveillance and privacy in relation to networked mobile cameras. For instance, for two decades the American computer scientist Steve Mann has worn custom made glasses with a hidden camera that records his everyday life and beams those images to a website. Mann's philosophy for "wearcam.org" reads like a combination of Dziga Vertov ("I am a camera"), Marshall McLuhan ("the internet as an extension of the nervous system") and Donna Haraway ("I am a photoborg entity"). Mann argues that the more cameras there are in the world, and the more democratically images are dispersed, the less chance there is of harmful corporate or state surveillance. Although flavored with a particularly American paranoia, his thesis of *sousveillance* or surveillance from below (an "inverse surveillance" involving community-based recording from first person perspectives) is highly relevant to current debates around privacy. More recently, Wafaa Bilal surgically implanted a camera in the back of his head for *The 3rd I* (2010–2011) project, streaming an image once a minute to a website that also showed his location via GPS, over a full year.[25] Bilal said he wanted to capture the mundane, but ironically his employer, New York University, required him to cover the camera while on campus. Such projects raise ethical issues that we are only starting to address.

What has become clear is that traditional theories of privacy are in need of renovation. As a historical and ideological product of the rise of bourgeois society after the Renaissance, privacy tends to be viewed as a right possessed by individuals—the right to be left alone, framed in exclusively individualistic terms.[26] But in relation to the ever-growing demand for more privacy, one always needs to ask: who benefits? Privacy laws often protect vested interests, and while it has become a question of identity management for ordinary citizens, advocates of new concepts like "privacy in public" must also recognize that individuals with cameras are hardly the most pressing threat.[27] In countries like Australia, surveillance is increasingly embedded and invisible, such as the data-mining of consumer transactions undertaken with the ambition of rendering desire more profitable via the formation of detailed personal consumption profiles. Such commercial activities should indeed be made more transparent—users should be the owners of their own data. However, on the other hand, there is no good reason why one's image in public should not remain just that, public.

Anxiety about being watched is nevertheless perfectly understandable in our culture, and many artists have mined this territory in their work. For instance, Rafael Lozano-Hemmer has explored tracking technologies in works such as *Surface Tension* (1993), an image of a giant human eye that follows the observers as they watch the image on screen. However, personal technologies like the mobile phone can also be put to social uses. In Lozano-Hemmer's large-scale interactive work *Amodal Suspension* (2003), SMS (short messaging service) messages were encoded as unique sequences of flashes by twenty robotically controlled searchlights, creating an extraordinary telegraphic spectacle over the city of Yamaguchi, Japan. The effect was to make public the most intimate of communications. From a more anthropological orientation, both as a researcher and as artist Larissa Hjorth has shown a long-standing interest in the gendered rituals and customization surrounding mobile phone use. In *CU: The Presence of Co-Presence* (2009), Hjorth made photographs corresponding to the emotional states of SMS messages, and their logic of the "ephemeral, compressed and abbreviated." These "fleeting moments of sadness, love and friendship"— and what the artist calls "contextual intimacy"—were then rendered public in an outdoor projection space in Melbourne.[28] In *Still Mobile* (2010) Hjorth used a mobile phone to take images of people taking pictures with their mobile phones, which she then abstracted into blurry digital images as if to suggest our subjective dissolution into the enfolded visual field.

In *Phone Camera Work* (2008), Patrick Pound exhibited a five-by-ten grid of blurry black-and-white A4 images pinned to the wall. Made with a mobile phone, the images look like the work of a spy or detective but are in fact cropped details from photographs in the daily newspaper—largely from the real estate pages. Consistent with his broader practice of presenting recurring themes in found photographs, for Pound the digital phone camera "is the perfect collecting machine . . . [that] can copy the world at will."[29] Pound revels in the redundancy of old technologies, using an old Motorola phone for its black-and-white possibilities, which he says approaches the low end technological look of surveillance footage and Atget's photographs all at once. He later emulated the look on the iPhone. Pound also used his phone to capture a sequence of images of a young photographer straining to make images of artworks at the National Gallery in Washington. He also merged the mobile phone with found photographs to mimic forensic photography in the series *Crime Scenes* (2011) taken from a collection of found images called *People that look dead but (probably) aren't*. Taking to its logical conclusion Walter Benjamin's claim in response to early twentieth century media of film and photography that "[e]very day the urge grows stronger to get hold of an object at very close range by way of its likeness, its reproduction," Pound's ingenious work with the phone camera reveals not reality but a hall of mirrors.[30]

Phone cameras can in fact be used not just to reproduce but also to enhance and augment our experience of the world. Taking advantage of the iPhone's GPS receiver, myriad "apps" have been built to facilitate navigation and social networking. Artists can work with, or even develop, such location-aware software to help reverse the increased personalization and commodification of urban space ordinarily brought about by such media. Thus, with their augmented reality artwork *Transumer* (2010), the Australian tactical media group pvi collective armed participants with modified iPhones and placed reality tags in the virtual landscape, mixing virtual imaging into the video stream of the mobile phone's camera in real time. The resulting "journey" aimed to reveal a hidden layer to the city, opening up possibilities for subverting official narratives of space. Thus updating Situationist techniques, the pvi collective seek "to devise fleeting moments of cultural intervention that are intended to temporarily transform the everyday . . . [and] establish a different kind of doing in these codified spaces."[31]

Smartphones, as Scott McQuire has observed, "can all too easily be aligned with the extension of commodity logic into the interstices of both public and private space."[32] As he suggests, the cultural problem is "not simply the exposure of the previously private", but rather the "failure to imagine new publics." What is needed—and what artists can provide, directly or indirectly—are moments of reflection and critique, to move beyond an unproductive structure of voyeurism and narcissism. Such a desire, to enable technologies for social ends, is also apparent in so-called "participatory sensing" projects.[33] Participatory sensing extends a history of citizen science projects, using embedded devices to capture data about oneself and one's community.[34] As Kate Shilton has argued, participatory sensing is an emerging form of mass data collection that may, under appropriate conditions, be conceived as a form of "empowering" self-surveillance—an opportunity for individuals and groups to "provide possibilities for self-exploration, community discovery, and new knowledge creation."[35] For instance, people can use a mobile phone's sensor capacities, such as the camera, to work together for specific purposes, from such simple projects as mapping local pollution to targeting and identifying invasive plant species in national parks or the best bicycle routes in that most notorious of car cities, Los Angeles. Such "reality mining," as it has been dubbed, operates as a kind of antidote to the corporate dominance of data mining. Indirectly, such projects remind us that simplistic concerns about the inappropriate use of camera phones too often reflect a misplaced anxiety about our powerlessness in a world of data and capital flows that are largely invisible.

In conclusion, the iPhone is part of a broader shift toward ubiquitous, convergent mobile media. Although the meaning of the iPhone for photography remains unclear, current evidence indicates that it represents the start of a new wave of computational photography, in which cameras are increasingly able to recognize and interpret scenes and the resulting images are used to "do" things. While cameras will still be used for more traditional purposes such as personal photography and photojournalism, these are now residual activities—as the nostalgic photo apps evocatively suggest. The more dramatic shifts exemplified by the iPhone begin with the popularization of photo-blogging but extend to the augmentation of every-day urban space and more specialized sensing activities such as tracking the user's eye movement across the phone's display as a means to activate applications. If the rumors are true that the iPhone 5 will include biometric face recognition log-in security using the front-facing camera, the fact that this technology has deep roots in nineteenth century appropriations of photography for criminological and surveillance purposes seems almost beside the point. Photography has long been involved in configuring the boundaries between private and public space, a task that is now combined with the mediation of individual identity in the psychic and social space of the mobile phone.

NOTES

1. This claim appears on Nokia's website in the "Camera phones backgrounder," http://www.nokia.com/NOKIA_COM_1/Microsites/Entry_Event/Materials/Camera_phones_backgrounder.pdf (accessed 20 June 2011).
2. Daniel Rubinstein and Katrina Sluis, "A life more photographic: mapping the networked image," *Photographies* 1(1), 2008: 9–28.
3. See Lisa Gye, "Picture this: the impact of mobile camera phones on personal photographic practices," *Continuum: Journal of Media & Cultural Studies* 21(2), 2007: 279–288; and Susan Murray, "Digital images, photo-sharing, and our shifting notions of everyday aesthetics," *Journal of Visual Culture* 7, no. 2, 2008: 147–153.
4. Daniel Rubenstein, "Cameraphone photography: the death of the camera and the arrival of visible speech," *The Issues in Contemporary Culture and Aesthetics* 1, 2005: 113–118.
5. Mikko Villi, *Visual Mobile Communication: Camera Phone Photo Messages as Ritual Communication and Mediated Presence* (Helsinki: Aalto University School of Art and Design, 2010), 150.
6. Anna Reading, "The mobile family gallery? Gender, memory and the cameraphone," *Trames: A Journal of the Humanities and Social Sciences* 12(3), 2008: 355–365.
7. With the iPhone 3GS in 2009, the camera gained another million pixels, autofocus and a "tap-to-focus" system—which links the focus control and auto-exposure system to a specific point in the image. It also gained the ability to shoot video. The iPhone 4 in 2010 included yet more megapixels, better low-light sensitivity, an LED flash (effectively a "torch," and often used as such), 720-pixel HD video recording, a front-facing camera and a digital zoom feature (a clever marketing term for "cropping"). iPhone 4 also has a wider angle of view—closer to that of a 28-mm on a 35-mm format camera, versus the equivalent of about 35 mm on older iPhone models. The iPhone 4S in 2011 offered further improvements in picture quality, effectively making the iPhone a reasonable substitute for a compact camera.
8. See, for instance, the "how-to" book by Stephanie Roberts, *The Art of iPhoneography: A Guide to Mobile Creativity* (New York: Pixiq, 2011).

9. Geoffrey Batchen, *Forget Me Not: Photography and Remembrance* (New York: Princeton Architectural Press, 2004).

10. Mia Fineman, "Phoning it in," unpublished conference paper presented at *The Versatile Image: Photography in the Era of Web 2.0*, University of Sunderland, 26 June 2011.

11. Gerard Goggin, *Global Mobile Media* (New York: Routledge, 2011), 48.

12. CBS launched EyeMobile for iPhone in 2008, allowing its users to submit photos and videos to the company's citizen journalism site: CBSEyeMobile.com. Several US cable networks had earlier launched similar efforts—CNN with iReport and FoxNews with uReport. Although not the first phone to be able to record video, Apple simplified the process with the iPhone 3Gs, with its built-in video recording power and ability to directly upload to YouTube. According to one report, in the six days following the 3Gs's launch in 2009, "YouTube reported an incredible 400% increase in mobile video uploads." See Kit Eaton, "Is the iPhone 3G S speeding citizen journalism forward?" *Fast Company* 26 June 2009, http://www.fastcompany.com/blog/kit-eaton/technomix/could-iphone-3g-s-change-citizen-journalism-forever (accessed 28 March 2011).

13. In 2009 American advertising photographer Chase Jarvis published a book called *The Best Camera is the One That's with You: iPhone Photography* (Berkeley: New Riders, 2009), linked to a website (http://thebestcamera.com) and an iPhone app. Aimed at encouraging aspiring commercial photographers, promising involvement in an "exciting new world of photography," his proselytizing and entrepreneurial project is arguably best understood as self-promotion for the "Chase Jarvis" brand.

14. Damon Winter, "Through my eye, not hipstamatic's," *The New York Times*, 11 February 2011 http://lens.blogs.nytimes.com/2011/02/11/through-my-eye-not-hipstamatics (accessed 30 March 2011).

15. Rosemary Neill, "Not a good look," *Weekend Australian Review* 2–3 October 2010: 5–8.

16. Daniel Palmer, "In naked repose: the face of candid portrait photography," *Angelaki* 16(1), 2011: 111–128.

17. See, for instance, Larissa Hjorth "Mobile spectres of intimacy: the gendered role of mobile technologies in love—past, present and future," in Rich Ling and Scott Campbell, eds., *The Mobile Communication Research Series: Volume II, Mobile Communication: Bringing Us Together or Tearing Us Apart?* (Edison, NJ: Transaction Books, 2011), 37–60; Larissa Hjorth, "Frames of discontent: social media, mobile intimacy and the boundaries of media practice," in Hill Koskela and John Macgregor Wise, eds., *The New Ecstasy of Communication: New Visualities, New Technologies* (Ashgate, 2012, forthcoming).

18. Ariella Azoulay, *The Civil Contract of Photography*, trans. Rela Mazali and Ruvik Danieli (New York: Zone Books, 2008).

19. Geoffrey Batchen, "Guilty pleasures," in Thomas Y. Levin, Ursula Frohne and Peter Weibel, eds., *Ctrl [Space]: Rhetorics of Surveillance from Bentham to Big Brother* (Cambridge, MA: MIT Press/ZKM Center for Art and Media, 2002), 447–459.

20. Eric Gordon and Adriana de Souza e Silva, *Net Locality: Why Location Matters in a Networked World* (Boston: Blackwell-Wiley, 2011).

21. *iPruitt* was exhibited in 2008 at the prestigious Gavin Browne's enterprise in New York. For a description of the exhibition, see the review by Katie Sonnenborn, "Rob Pruitt," *Frieze*, January-February 2009: 153–154.

22. Fineman, "Phoning it in."

23. Joel Sternfeld, *iDubai* (Göttingen: Steidl, 2010).

24. Joel Sternfeld, "Joel Sternfeld: iDubai" (interview), *Daylight Magazine*, November 2010 http://www.daylightmagazine.org/podcast/november2010-0 (accessed 19 April 2011).
25. http://www.3rdi.me (accessed 20 March 2011).
26. Jessica Whyte, "Criminalising 'camera fiends': photography restrictions in the age of digital reproduction," *Australian Feminist Law Journal* 31, 2009: 99–120.
27. Helen Nissenbaum, "Protecting privacy in an information age: the problem of privacy in public," *Law and Philosophy* 17, 1998: 559–596.
28. Larissa Hjorth, Centre for Contemporary Photography (CCP), Melbourne, July 2009.
29. Larissa Hjorth, "Photoshifting: art practice, camera phones and social media," *Photofile* 89 2010: 32–37.
30. Walter Benjamin, "The work of art in the age of mechanical reproduction," in Hannah Arendt, ed., trans. Harry Zohn, *Illuminations* (New York: Schocken, 1969), 219–253.
31. The Australian Council, "Augmented reality intervention into urban spaces," http://www.australiacouncil.gov.au/artforms/inter-arts/news_items/transumer (accessed 10 April 2011).
32. Scott McQuire, *The Media City: Media, Architecture and Urban Space* (London: Sage, 2008), 204.
33. See, for instance, the work carried about by scholars at University of California, Los Angeles's Center for Embedded Networked Sensing, including Jeffrey Goldman, Katie Shilton, Jeff Burke, Deborah Estrin, Mark Hansen, Nithya Ramanathan, Sasank Reddy, Vids Samanta, Mani Srivastava, and Ruth West, "Participatory sensing: a citizen-powered approach to illuminating the patterns that shape our world," Woodrow Wilson Center for International Scholars, May 2009, http://wilsoncenter.org/topics/docs/participatory_sensing.pdf (accessed 10 April 2011).
34. Kate Shilton, "Participatory sensing: building empowering surveillance," *Surveillance & Society* 8(2), 2010: 131–150.
35. Ibid.

7 Between Image and Information
The iPhone Camera in the History of Photography

Chris Chesher

Raise the iPhone with its back pointing toward your subject.
Tap and hold the camera button on the touchscreen.
Release the button. Hear the familiar shutter release sound effect,
and watch the shutter animation.

INTRODUCTION

When the iPhone camera snaps a scene, as above, it is not only optics and a photo-sensitive surface that determine the outcomes, as with conventional cameras. The device can perform many combinations of digital operations, including analyzing the image data, performing algorithmic changes, connecting to other data spaces and storing image files. For example, the iPhone might return a dramatically processed photographic image, create an office document or use the image as pure information. This wide range of potential operations is courtesy of the iPhone's key software innovation: apps, bundles of meaning and functionality each marked by its own distinctive name and icon and available for sale in the App Store. Any app that uses the iPhone's camera becomes an interface between user events of photography and a particular set of possible visual and informational processes. Despite having a camera with major technical limitations, the iPhone has become a disruptive technology in amateur photography. I will argue that it opens onto its own "Universes of reference"—technical, aesthetic, subjective and instrumental—and has established new aggressive economic models.

This chapter explores how the iPhone, supported by the distributed app development community that feeds Apple's App Store, has participated in a minor re-invention of the amateur camera. By turning to software, the iPhone camera opens onto heterogeneous new existential territories, extending beyond conventional image capture and into instant digital manipulation, pattern recognition, augmented reality and even medical monitoring. Each of these possible pathways for the image reveals the iPhone's affiliation with a new "Universe of reference," beyond conventional photography. The Universe of reference is a concept from Félix Guattari's essay on technology "Machinic Heterogenesis"[1] that I will use to examine historical changes in the technosocial configurations of mass amateur photography. The iPhone Universe is connected much more immediately to computing traditions (information, databases, algorithmic transformations) than it is to film, digital and mobile cameras. The amateur image becomes immediately transformable, transmissible, transcodable, and taggable in real time. At the same time, the imaging practices using iOS cameras and image apps become subject to Apple's tight monopoly control over the App Store.

When the iPhone arrived in 2007, it began drawing on existing investments in media culture. Media consumption had long been an important part of everyday life. Most people in the developed world already owned and valued their media technologies (DVD players, televisions, computers, cameras) and media content collections (DVDs, music, software and so on). The new artifact established relationships with older media, through a range of mechanisms and strategies. Apple's successes with the iPod, the Macintosh and iTunes were strategic assets Apple controlled, but this was not all. The wider mediascape was a virtual resource that Apple and its users could exploit symbolically and materially.

Even before the iPhone was released, Apple began seeding popular expectations. A key strategy was to reference existing media cultures. For example, the company booked a TV advertisement to screen during the TV coverage of the 2007 Academy Awards. The ad featured fast-cut scenes from many movies from different eras and genres. In each, a recognizable actor answers a phone.[2] In this barrage of "Hellos," characters from movie history seem to greet the iPhone. The ad draws together three cultures of technology: the audiovisual history of cinema, telecommunications history of the phone and consumer computing. The ad helps call into existence what Guattari calls a new "Universe of reference": The iPhone Universe of reference.

The iPhone Universe of reference is best explored by tracing the extended connections between actual and virtual components. Each of iPhone's technical components—video playback, phone calls, music, web browsing, global positioning, image capture—establishes certain potentials for mediated events. The mix of components establishes a virtual field of connections that are real, but not yet actual. Using Guattari's schema, the iPhone Universe extends well beyond its technical parts, to include dynamically interconnected and extended social, mental and abstract components: material and energy components; semiotic, diagrammatic and algorithmic components (plans, formulas, equations and calculations that lead to the fabrication of the machine); components of organs, influx and humors of the human body; individual and collective mental representations and information; investments of desiring machines producing a subjectivity adjacent to these components; and abstract machines installing themselves transversally to the machinic levels previously considered (material, cognitive, affective and social).[3]

The last point is key, as components are connected cross-wise by what Guattari refers to as "abstract machines," the underlying material and cultural forces and resources that drive regularities and innovations.

THE UNIVERSES OF REFERENCE OF THE iPHONE'S CAMERA

This chapter now returns to the Universe of reference of mass amateur photography, and the disruptive influence that the iPhone appears to be opening up. Through human history, the dominant materials and methods in a culture have helped define the limits and possibilities for an era. As changes in materials arise, new abstract machines unfold, with their own Universes of reference, colliding and combining with older Universes. The processes of combination are non-linear and chaotic, and their emergent morphologies reconfigure how events unfold. Guattari refers to these generative assemblages, formed by mixing a diverse range of elements, as heterogeneous machinic Universes:

> It is at the intersection of heterogeneous machinic Universes, of different dimensions and with unfamiliar ontological textures, radical innovations and once forgotten, then reactivated, ancestral machinic lines, that the movement of history singularizes itself.[4]

The iPhone camera mobilizes longstanding subjective impulses for making images, common not only to Kodak, but also to the motivations for cave drawings, oil paintings and daguerreotypes (a long-term, constantly changing abstract machine). However, iPhone users take up these affective forces and aesthetic values in different ways, with different materials, forming different connections. The iPhone camera moves into significantly different existential territories from its recent progenitor, the film camera. It is a disruptive technology not only because it has new technical features—low-power complementary metal oxide semiconductor (CMOS) image sensors, solid-state data storage, multimedia and telecommunications capacities. It also enters into relationships with new forms of creative production and circulation, including the sometimes heavy-handed influence of Apple's aspiration to tie its own devices and processes into new forms of strategic control over markets. The differences between chemical, digital mobile and smart cameras must be understood within interconnected histories of image cultures, habits of consumption, systems of production, image manipulation practices and strategies for market domination. At each change in the table below, new patterns of use were enhanced (+) or diminished (–) in comparison with previous practices.

	Film camera	Digital camera	Feature phone	iPhone / smart
Image capture practices	Hobbyists; family; tourism	+ cheaper = more shots + Immediate preview	+ ubiquitous availability; + transmission	+ touch screen as viewfinder
Subject matter	Normal growth and change in family	little change	+ pragmatic uses + more personal images + on-the-spot	+ variable image modalities
Manipulating images	Expensive specialist processing	+ computer manipulation - evidential value	- image quality	+ in-device digital processing, interpretation
Ritual viewing	View in a separate space and time	+ new modes of exhibition: + online photo sharing & discussion; + digital photo frames	+ mobile display + always available	+ integration with social media
Economic model	Kodak monopoly over paper and film	Reusable media: memory cards; Printers & inks	+ telecommunications subscriptions & charging	+ Apple App store + App market (ads; in-app purchases)

Figure 7.1   Usage patterns in amateur photography from film to iPhone.

CONTINUITIES AND RUPTURES IN IMAGE-MAKING SINCE 1899

As the table above shows, the Universes of reference for iPhones and other smartphones (such as Android and Windows phones) diverge in significant ways from conventional cameras in the three previously dominant mass imaging traditions: film cameras (1899–2000), digital cameras (1986–) and "feature" phones[5] with cameras (1992–). Each of these popular camera traditions supports not only different patterns of image capture and modes of manipulation, but different economic models. The iPhone camera, compared with most dedicated cameras, is inferior in optics, image quality and ergonomics from dedicated cameras. It is also relatively expensive. However, like all mobile phone cameras, it is always at hand for users. Unlike the other categories, though, the iPhone camera has apps that allow it to be used in many different ways beyond conventional image capture.

In spite of the differences between analogue, digital, feature phone and iPhone cameras, there are many continuities in the cultural practices associated with image making and consumption. The rituals of taking particular genres of photos—the portrait, the group photo, the candid snap, the art photo and so on—are in many ways common across these technologies, even if they change in ways discussed below. Themes such as identity, family, friendship, visual experimentation and so on remain powerful throughout recent image making. Historical changes in media have contributed to differences in perceived meanings, such as when digital manipulation complicated the evidential value of the photographic image.[6] The iPhone extends a long story about how changes in equipment, materials and marketing contribute to establishing new user groups and new industries of the captured image (see Chapter 6 in this collection).

KODAK'S LONG MOMENT

The invention of mass amateur photography is often dated to 1900, when Kodak introduced the "Brownie" camera. More than any object, the box Brownie is the artifact metonymic of the Kodak Universe of reference. Kodak's Universe was perhaps the first example of user-generated content: offering people the capacity to use high technology to create their own images, but tightly controlling the production process. The famous slogan "You push the button, we do the rest" aimed to alleviate the anxiety of those who feared photography's technical complexity. It also symbolically diminished the value of the user's labor, which created the value in the photo. Traces of this labor are apparent in the visual cultural legacy of billions of photographs—families, sports, public events, portraits and experiments—that betray the dominance of the Kodak Universe of reference. Kodak was synonymous with the amateur image-making industry for over a century.

However, stabilizing this model for photography had been difficult. The Kodak moment arrived only after decades of technical, legal and business struggles and innovations. In the mid-nineteenth century, photography was expensive and confined to specialists.[7] Early chemical photography required substantial skill in handling complex and time-sensitive processes. As Jenkins notes, "Every photographer served not only as the camera operator but also as a decentralized handicraft producer of photosensitive materials and of finished positive prints."[8] Gradually, by creating multiple innovations—more stable chemicals, coated roll film rather than plates, punched holes—the Kodak Universe of reference became a more durable assemblage. Another key decision was to move from focusing on professional photographers, who wouldn't compromise their plate-making craft by using roll film, to a new market for amateur photographers:

> Eastman conceived of utilizing these resources to transform the roll film system intended for the professional to an amateur system of photography consisting of a simple-to-operate film camera, stripping film, and a factory service for developing and printing the delicate and hard-to-operate film.[9]

In analyzing Jenkins' history of Kodak, Bruno Latour,[10] a pioneer in the social studies of technology tradition Actor Network Theory (ANT), observes that changes in technical image making were simultaneous with the formation of new social groups, meanings, behaviors and reconfigured cultural practices. The lines of causation between these are always multi-directional. Influence never goes *only* from technological to social change, nor the reverse. Using the ANT approach (which shares with Guattari's Universes and machines the emphasis on diverse interconnected components), humans and non-humans are treated symmetrically as *actors*. The Kodak brand, the Brownie box, the snapshot, the amateur photographer and photographic subjects who know how to pose are all actors reconfigured at each singular historical intersection through processes of *translation*. Imagining the configurations of people and technologies as geometrical entities, Latour argues that the ways that humans (proto-photographers) and non-humans (chemicals, film etc.) might be organized remained undecided for most of the time during which the model of mass amateur photography was forming.

> What we observe is *a group of variable geometry entering into a relationship with an object of variable geometry*. Both get transformed. We observe a process of translation—not one of reception, rejection, resistance, or acceptance.[11]

Once "Kodak culture" prevailed after 1899, amateur photography in "home mode" began to document particular Universes of reference: the "normal" growth and change in the home, particularly in children.[12] The camera typically opened onto Universes of the home and family. Kodak's promotional campaigns emphasized at various times the fun of photography, the best uses of the cameras for children, capturing candid moments and the importance of domestic memories.[13] Beyond the home, uses of cameras became essential in tourism, photographic art, journalistic reportage and so on.

Over the years, some different camera practices emerged, such as when Eastman Kodak's 1923 16-mm movie format made home movies possible. In 1948 Polaroid Corporation released a new camera that instantly processed its own photos, establishing a parallel minor history in photography.[14] Sony and JVC introduced domestic video cameras in the 1980s. Each of these image assemblages opened onto slightly different Universes of images, extended and constrained variously by image quality, costs, temporality, portability and so on.

In the film era, critical commentary placed a heavy significance on the psychological and cultural values of photography. Photography is an important theme for writers as influential as Sigmund Freud,[15] Walter Benjamin,[16] Roland Barthes[17] and Jacques Derrida.[18] It provided them with a powerful analogy for wider cultural themes: authenticity, fleetingness of events, the fragility of memory and the inevitability of death.[19] Certainly these associations have genesis in the uncanny resemblance between the living face and the photographic image. However, these effects were enhanced by the materiality of analogue photographic production: the dark cavity in the camera that cannot be exposed to light; the routines of loading and unloading expensive film reels; the slow chemical and light enlargement rituals that gradually reveal images; the manipulations in the darkroom relying directly upon interrupting light in the exposure process; and the shiny, precious prints that emerge. As captured images became cheaper, faster and more easily reproducible, the regimes of value in the Universe of photography change.

DIGITAL CAMERAS AND THE GEOMETRIES OF DIRECT DISPLACEMENT

During the 1990s, a new kind of camera, associated with the Universe of reference of digital electronics, moved in to challenge the cultural geometry that film photography had dominated. The retail release of digital cameras followed large investments during the 1980s from a diverse range of companies working on standards, techniques and components for digital cameras. They included builders of film cameras (Nikon, Canon), film stock (Fuji, Kodak) and electronics and computing (Casio, Apple, Dycam, Logitech and Sony). Those companies with investment in film developed digital devices with some ambivalence: they were contributing to the death of their core business.

Once standards were settled, digital camera sales took off, overtaking film camera sales in 2003. After less than a decade, the digital image relegated film to a specialist or nostalgic technology. Digital cameras almost directly *replaced* film cameras, as they took over and extended most of the cultural functions of film cameras.[20] Manufacturers built cameras along very similar lines to film predecessors: cheap snappy cameras, mid-range zoom cameras and digital SLRs. But digital was cheaper to shoot, with immediate preview and greater capacity to alter, select, transform, transmit and print the images once back at the personal computer.

At the same time, the meanings of digital images were subtly different. The fetish of the indexical link between the light of the scene, the light of the darkroom, and the light reflecting off the print seemed to be broken.[21] The digital image is apparently more subject to retouching in the Universe of reference of pixels in digital storage. More importantly, the relative cheapness, flexibility and accessibility of digital cameras and images help demystify the digital image. There is an even smaller auratic,[22] spatial, temporal or economic gap between the event of exposure and the moment that the image is revealed (not even the tantalizing seconds of the Polaroid). Images can easily be selected, edited and sent electronically to multiple locations.

The new digital paradigm has major implications for Kodak Universe incumbents, particularly those selling consumables such as film and photosensitive paper.[23] They find their markets literally drying up. On the other hand, companies making ink-jet printers and consumables prosper, pursuing market strategies of controlling aftermarket consumables that is in many ways very similar to their predecessors.[24]

FEATURE PHONES AND THE UNIVERSE OF CAMERA UBIQUITY

When it became possible to include small digital cameras in phones, the artifact that emerged was quite unlike previous domestic cameras. The first mobile phone featuring a rudimentary 0.1 megapixel camera was the Sharp J-SH04, released only in Japan in 2000. In November 2001 Nokia released the model 7650, the first GSM (Global System for Mobile Communications) camera phone.[25] Within five years, almost all phones featured small, cheap CMOS cameras. These cameras tended not to compete directly with family photography, functioning instead as mobile personal imaging devices. With limitations in lens-depth, sensor size, limited storage and low cost electronics, their image quality was almost universally poor, so they never threatened to displace the full technical or cultural functions of the film or digital camera.

However, it was the omnipresence of these cameras, which went everywhere people took their phones, that was often perceived as intrusiveness. Stories of regulatory reactions began appearing around the world. Saudi Arabia reportedly banned these devices on the pretext that "men were reported to be using the devices to photograph women secretly."[26] An article in the *Beijing Review*[27] notes that bathhouses banned camera phones; and more ominously, suggest such devices may be illegal for security reasons: "Bureau of State Security in Fuzhou City, Fujian Province, said that the Law of State Security defines the special spy equipment as equipment with a built-in bug or camera."[28]

Along with regulatory reactions, the ubiquitousness of mobile phone cameras has other implications. Their omnipresence contributed to changes in typical image genres. Any random moment could be captured, and not only those pictures staged when a camera was present. Every possible image (and movie), from parties and domestic scenes to amateur reportage of tragedies—earthquakes, tsunamis, terrorist attacks and so on—would be captured with these shaky, lightweight artifacts. As it became possible to transmit images directly from the point of capture, the liveness of the mobile image recovered some of the evidential value lost to digital manipulation. The practice of "moblogging"—telling an ongoing story through images uploaded from the phone—was supported in the design and promotion for feature phones such as the Nokia 7710,[29] although the practice of telling stories through uploaded images did not become widespread until smart phone cameras were paired with social media.

Mobile phone images tend to have a stronger association with individuals than with families. Phones typically (but not always) belong to a single person; therefore, the images that accumulate in the camera relate directly with that one user's everyday point of view and experience. With internal storage, the phone itself became a personal image collection, implicated in mediating the user's sense of identity.

> A photographic archive of memories, a mobile archive; always within easy reach, something to look at again and again, when feeling nostalgic, or just to pass an interstitial moment in one's daily routine.[30]

The feature phone's camera and screen became a self-contained simple mobile imaging apparatus, constantly at hand and largely self-contained. While it was *possible* to upload images to a computer, or send images to others over mobile messaging service (MMS), most users shared images with others using the phone's screen rather than electronically.[31]

iPHONE APPS TRANSFORM, TRANSLATE AND TRANSMIT IMAGES

When the first model of the iPhone was released in 2007, the camera seemed unlikely to change the world. It was relatively low resolution (two megapixels), fixed focus and the default app couldn't even zoom the image. It seemed to be just another phone camera. However, the iPhone was establishing new Universes of reference by bringing very different technical, cultural and institutional entities into its orbit. It could connect directly to the internet, and not to a telecommunications company's "walled garden." As a network-attached small mobile computer with a large touchscreen, it was a platform that opened a camera onto new existential territories.

After the App Store was launched in July 2008, a number of developers introduced apps that used the camera in ways other than taking standard snaps. The App Store, based on Apple's iTunes Music store, the most successful online download music service since 2003, created a very competitive domain. As the number of apps exploded to over 300 thousand available,[32] developers increasingly had to make their apps really stand out. Meanwhile, newer models of the iPhone featured incremental improvements in the camera: especially iPhone 4 with higher resolution and superior touch-to-focus lens. The sheer number of iPhones sold—over 100 million in the first four years—helped make this unlikely device top the "most popular cameras" list on the image-sharing site Flickr.[33] The App Store's significance grew with this popularity, as the monopoly channel for content and apps for iOS devices: iPhone, iPod Touch and iPad. After 2009, developers could sell additional content or features as "in-app purchases." The App Store became a center of market power and regulatory control: Apple vets all apps, negotiates content rights and takes 30 percent cut from all income from apps and content. In 2011 the App Store passed 10 billion downloads, surpassing the number of songs ever sold through the iTunes Music store.[34]

What distinguishes the iPhone as a camera is its capacity to perform real-time digital transformations, translations and transmissions on mobile amateur images. The camera does more than capture images. In a minor way, it enters the realms of media production, information and deixis (it carries information about person, direction, time etc. with its images). By 2010, taking images with the iPhone had become identified as the distinctive practice of iPhonography.[35] While iPhone's basic camera opens onto similar Universes of reference as the Kodak, digital and mobile image Universes, with apps it reconfigures, or even gets rid of images altogether.

In addition, iPhones and other smart phones mediate significantly different practices in making and consuming images. The remainder of this chapter examines a number of iPhone apps that use the camera in new ways. It begins with apps that use software to transform images to open them onto visual Universes of reference beyond those established in the photographic event: *Hipstamatic*, which generates nostalgic images by simulating the features and distortions of old cameras, lenses, films and flash, and *Camscanner+*, which functions as a scanner, cropping the images of documents and adjusting for distortions.

Many apps take the camera beyond its photographic heritage to use it as a data input device, collecting information instead of making conventional photos. *Google Goggles* takes snaps and matches distinctive features in the images to its own databases, such as English or non-English text, well-known landmarks, artworks, logos, QR codes and bar codes. Instead of images, it presents translated information. Augmented reality apps such as *layar* also deal in information, overlaying deictic (pointing and indicating) labels onto the viewfinder image without capturing any shots. An even more extreme example of the changing the role of the camera is *Heart Rate Monitor*, which asks users to cover the lens with their fingers to measure their heartbeat.

SIMULATING THE NOSTALGIC IMAGE

Hipstamatic is a camera app that simulates extremes of photographic imaging. When a photo breaks down over time or is manipulated in the darkroom or is captured with an unconventional camera such as the Russian LOMO (which became popular in the 1990s) or the Polaroid, it can no longer be read as a realistic photo. These images tend to create an effect of nostalgia, other-worldliness or coolness. *Hipstamatic* is an app for autonostalgia, opening onto what might be described as arty, sophisticated, underground Universes of reference in visual culture. It transforms the simple camera phone snapshots into nostalgic *photographs* belonging to another time and place.

Users use the app's retro-styled interface to switch (virtual) lenses, films and flash. These choices alter the settings for post-processing of the image. They change the contrast, adjust color caste, add borders or introduce noise and distortion. These effects stage dramatic digital manipulations of the image that could previously only be performed on a computer, or in a more limited way with some digital cameras with digital signal processors.

The choices of virtual equipment are documented in the website for the app with evocative descriptions. According to the *Hipstamatic Field Guide*,[36]

> The John S Lens is a good all-round lens, and works well in low-light, bright-light, and just about any other scenario. It creates a vignette effect with saturation in cool tones.

Or for a different effect,

> Domo arigato, this lens is better than winning the lotto. Add a burst of super happy robot love to your images. Imported from Japan . . . Lens Qualities: Strong purple and teal shade added to a center focus star burst pattern.

The success of *Hipstamatic* is in packaging and commoditizing digital image effects as a toy-like virtual camera, with add-on features that must be bought with in-app purchases. The settings invoke both temporal and cultural distance. In combination with these lenses the photographer can choose a film called "Ina 1969" (adding rounded corners and a textured border) and the "Dreampop" flash (creating random color changes). *Hipstamatic* makes image capture deliberately difficult, initially offering only a small "viewfinder," moving the framing of the image at random, and imposing a long delay while "processing" the image (later versions allowed queuing). Some users see that all these limitations actually contribute to the aesthetic experience of using the app.[37]

Instagram is another app that trades in instant nostalgia. It positions itself as a hybrid of the polaroid and the telegram: a neo-retro device that makes instant images transmissible anywhere. Users capture an image and choose image filters such as Lomo-fi, Earlybird, Toaster and Poprocket to enhance their visual impact. Both *Hipstamatic* and *Instagram* lower the image's semiotic modality[38] so that it is read less as a naturalistic photographic image, and becomes "more than real."[39] For Kress and Van Leeuwen, the meanings of changes in visual modality such as exaggerations of color, enhanced sharpness and other effects are to perform social differentiations: "Modality both realises and produces social affinity, through aligning the viewer . . . with certain forms of representation, namely those with which the artist aligns himself or herself, and not with others."[40] Choices of images and modalities can align with, and reinforce, social Universes of reference. Images manipulated in ways that appeal to an in-group (gender, ethnicity, fan, and subculture) can help define and confirm the in-groups and out-groups.

If the iPhonographer's choice to transform an image in a particular way is implicitly social, *Instagram* accelerates the infrastructure for sociality by giving users the opportunity to share images online. It is the norm for users of social networking services to manage accounts in a number of different social media simultaneously, thereby maintaining connections with multiple overlapping peer groups. The instant photographer can immediately send transformed images to any of a number of social media sites: social networking, blogging and photo sharing. Transmitting images compresses the temporal gap between production and consumption, and into the timeframe of tactical media. With tactical media, the recipients of images can respond in real time, and even react in the manner of flash-mobs or during political events[41] such as in Iran in 2009 and Egypt, Bahrain and Libya in 2011. More trivially, a Facebook user on a day trip may find several comments on an uploaded image even before they return home.

ENTERING THE OFFICE

The app *Camscanner+* opens onto the very different existential territory of the office.[42] To *scan* a document with *Camscanner+*, the user simply points the camera at a document and frames it as tightly as possible. The app waits until it senses there is not too much camera movement, and only then takes the image. It presents the user with an image of the document, and four handles to mark its corners of the document. The user moves these handles into position, and taps on the tick button to proceed. An animation shows the app correcting the parallax, transforming the document with "Magic Color" and cropping the image into a clean, symmetrical scan. The user can upload the image as a *scanned* file over Wi-Fi or 3G, or send it over e-mail as a PDF.

At least at the levels of features and aesthetics, *Camscanner+* supports norms of accountability, transparency and standardization characteristic of the professional Universe of reference of the office. Guattari's ethics distinguishes between structures that seek permanence, and machines that constantly change and produce the new. This app embodies an alignment with structure and repetition rather than with productive disequilibrium. "(I)t puts into play a concept of totalisation that it itself masters. It is occupied by inputs and outputs whose purpose is to make the structure function according to the principle of eternal return."[43] In its modes of expression, and workflow, *Camscanner+* aspires to generate homogenized collections for integration into institutions.

GOGGLE RECOGNITIONS

Unlike the *Hipstamatic, Instagram* and *Camscanner+, Google Goggles* uses images as a visual form of search query. With *Goggles*, the image is only a disposable, intermediate step on the way to information. *Goggles* is currently an experiment (in "Google Labs") that runs within the iPhone Google app (also available on Google's Android platform). The first part of its operation is something like any camera app: the user takes a snapshot. Once the image is captured, though, the interface shows a swarm of animated colored boxes floating up across the image. The camera transmits the image information to a Google server to match it against a huge image database. Once the processing is complete, *Goggles* shows the image with colored boxes overlaid, and an index of image query results underneath. Each color marks that a section has been interpreted according to a different semiotic mode: Text, Logo, Landmark, Book, Music, Celebrity, Product, Movie and Barcodes. *Google Goggles* reads composite images, such as a book cover, according to multiple modalities and codes. For example, the type in the title may be recognized as text, which is converted more or less successfully using optical character recognition. If the book includes an image of a well-known painting, that might be recognized as an artwork. The whole cover may also be recognized as a book, in which case *Goggles* presents links to the publisher and booksellers.

In *Goggles*' eyes, users' images potentially have meanings at multiple levels. However, the image itself must feature particular visual modalities: typeset text; recognizable images; barcodes and so on. In the process of capturing the image, the user needs to compose for recognition by Google, and not the usual photographic values of visual composition. *Goggles* works because everyday life is filled with the same images— traces of centuries of mass production, standardization and celebrity. Those who can capture an image are likely to be physically close to the object or brand for which they are searching. For Google, *Goggles* promises to become a location-based advertising platform,[44] extending into the domain of images the successful model of "Google Adwords," which connects natural language terms with target markets.

The most controversial possibility for *Google Goggles* is its unimplemented capacity to recognize faces. This would allow someone to take a snap of someone passing by and submit it to match against Google's databases.[45] When Google introduced *Goggles* in 2009, they decided not to enable the face-recognition features, citing privacy concerns. Large-scale face recognition is typically used in policing and border security applications, but there is not yet an app for that on the iPhone. However, it seems this technology is irresistible to some companies, as in 2011 social media site Facebook chose to enable a feature called "Tag Suggestions," which recognizes faces in uploaded photos.[46]

Google Goggles' ways of seeing are significantly different from human perspectives on the photograph. It reads images in according to what Guattari calls the "a-signifying semiotic."[47] The iPhone reads symbols and images directly as information, in addition to their conventional meanings as images. In classical signification, the signifier (the word "camera," or the image in a photograph) denotes a signified (the idea of a camera, or the impression of the photographed scene) through acts of interpretation. When an ATM reads the magnetic stripe on a credit card, there is a process of a-signifying semiotics, according to Guattari. *Goggles* is an extended mobilization of a-signifying semiotics in which the camera reads the scene as if it were a magnetic strip. Using this app, images no longer primarily function as visual signs. Instead, they open up new existential territories of real time translation by recognizing distinctive shapes in the database, interpreting them according to the different image types, and connecting with internet links. The images operate as:

> "point-signs": these on one hand belong to the semiotic order and on the other intervene directly in a series of material machinic processes . . . the a-signifying semiotic figures don't simply secrete significations. They give out stop and start orders but above all activate the "bring into being" of ontological Universes.[48]

Google Goggles flips visual Universes into informational Universes. The algorithms recognize features in the image against a database of images. This technique has been used for some time for documents with optical character recognition. Currently, though, *Goggles* does not function consistently well. When it fails to find a hit, it will often return a page full of suggested images. The peculiar outcomes suggest a completely alien intelligence is at work. For example, in capturing image of the view out my study window, *Goggles* returns no close matches, but suggests images of: crab fishing in Alaska; a public toilet in Eisenach, Germany; and Armored Core fighting robot; and a sculpture of Padre Cicero in a box in Argentina.

iPHONE DEIXIS: AUGMENTED REALITY

Another genre of camera app flips information into the user's field of vision: augmented reality. Apps such as *layar, Panoramia* and *Localscope* present the standard "viewfinder" perspective on a scene, sometimes overlaying a grid to suggest perspective. The app uses GPS to determine the user's position, and the compass and gyroscope to calculate which direction the user is facing. It matches this information about the user's position with entries about locations in databases such as search engines, Wikipedia, social media sites, and even custom sites for cultural institutions, city walks, and art projects.

The augmented reality regime of vision has less connection with Kodak culture and more resemblance to a gun sight or a head-up display in a modern fighter jet. It visually interprets the surrounding space with a view to action. It is a hybrid immediate and hypermediated image[49] presenting an immediate perspective through the frame of view-finder, combined with layers of information graphics (such as on the gun sight's aim point, shutter speed and other information). A different style of augmented reality app recognizes particular patterns on camera, and change the image itself. For example, *Word Lens* attempts to recognize text in another language, and translate it in place. It erases the unfamiliar text and overlays it with the familiar.

iPhone AR typically points towards opportunities for consumption: the closest cafes, bars, clubs, hotels and so on. It imposes a privileged set of meanings onto the surrounding space, with the label literally covering the site itself in the image. Its meanings are spatially deictic—the display is meaningful only in relation to the spatial context of the user. The deictic relationship is both bodily and informational: a miniature radar map read-out gives directional cues that the user follows by spinning around to discover the labels floating in space. The labels give the names and details about locations and a measurement of how far it is from the user.

SEEING THE HEARTBEAT

The final example of a re-invention of the camera in the iPhone entirely abandons its relationship to photography and affirms its function as a digital input device. Rather than surveying the surroundings, it monitors the body of the user. Users launch the *Instant Heart Rate* app and place their finger over the camera. The camera recognizes changes in the color of the skin as the blood becomes more or less saturated with oxygen. The iPhone becomes a kind of medical device or fitness monitor, presenting a metallic circular display with a numerical read out of the user's heart rate, and a graph visualizing the beat. It plays a loud beep on each beat. It also stores and analyses the performance, tracking the heart rate recovery after exercise and keeping a record over weeks and months.

Instant Heart Rate (IHR) turns the iPhone into a personal health-monitoring tool. It is among many commoditized devices that present the user with affirmations or warnings, and records whether they are fulfilling the societal expectations and norms around good health. The *IHR* encourages users to upload their heart rate information to social media sites Facebook and Twitter. The information *IHR* collects from the body is not private, like medical records, but public text with potential to generate cultural capital, documenting how fit the user is. Just as iPhones are often used to upload images of the self, family and friends captured by the camera, so the camera captures information about the user's body, and shares it online with social media friends. As a toy piece of medical or sporting equipment, *IHR* gives users a capacity to capture patterns in the body as digital information, but on their own terms. These kinds of device return to users a sense of control that is arguably lost in medical bureaucracy.

iPHONE CAMERAS MEDIATE NEW TECHNIQUES AND ECONOMIC MODELS

In consort with apps, the camera embedded in the iPhone is consistently becoming something other than a camera. *Hipstamatic* not only captures images but also digitally processes them immediately to mimic old film and other special effects. *Instagram* transmits its images to social media sites, both shortening the time between shutter release and public exposure, and transforming the modality of the image to provide apparent distance. *Camscanner+* transforms images of documents into documents. *Google Goggles* uses the camera as a sensor opening onto worlds of information, beyond images, recognizing visual patterns as search queries. Augmented reality apps align abstracted spatial information and GPS with the world as seen through the iPhone as viewfinder. Heart rate apps turn the camera into a medical sensor.

Each of these apps uses software to transform the relationship between camera and scene. Each uses different codes and different relationships to the body. Users must adjust to the app's Universe of reference, and frame the image for its conventions. Apps that perform artistic transformations are closer to conventional image capture. The photographer frames the scene as portrait, two-shot, wide-shot or landscape; adjusts the lighting; and anticipates the expressions and movements of the subjects.

With the other apps, the subject matter and method of photographing are not the same as conventional photography. Capturing an image for the *Scanner+* app, the photographer must make sure the document is revealed in even light, and is as straight and as tightly framed as possible. *Google Goggles* users learn to spot features in the environment that the app can recognize: logos, landmarks, book, CD and DVD covers and so on. The augmented reality apps are more like information displays, with no capacity to capture the image. Users constantly survey the scene with their hand up, eyes on the display. Finally, the image on the screen is not displayed at all on the *Instant Heart Rate* app, as the camera has become a bodily monitor.

What the apps share is the iPhone platform and the site of purchase: Apple's App Store. This market rewards third-party developers for their innovation and differentiation, while securing Apple as the exclusive provider of apps. Apps that simply duplicate existing functionalities are less likely to be bought. This has led to some diversity in photo apps: offering transformations such as color accent (picking a single color to accentuate against black and white); converting an image into a cartoon (*ToonPAINT* or *Comic Touch*); stitching together a panorama (*Photosynth* or *Autostitch Panorama*); defocusing a section of the image to simulate depth of field (*Tiltshift*); or adding cats into the picture (*Catpaint*). It seems likely that future models of higher end cameras will feature touch screen interfaces and App Store connectivity similar to iOS or Android.

In some ways, iPhone's transformation of the camera into a multi-purpose input device is in line with many other digital camera applications beyond amateur photography, from the battlefield to security cameras. iPhone apps draw directly and indirectly upon many decades of university, military and commercial research into image analysis, image transformation character recognition, head-up displays, and so on. The difference with iPhone cameras is the powerful processor embedded with the mobile personal camera, and the App Store with exclusive access to the market for the commoditized specialized software packages running on iOS, each of which opens onto a different Universe of reference.

This iPhone Universe of images is only new, and so far there are only early indications of its characteristic practices and genres. Undoubtedly many current image styles will quickly become dated and cliché, as their modalities become even more distant. Others practices may develop further sophistication as users and groups take up and develop techniques that exploit complex combinations of software and image capture. Apple's business model, controlling hardware, content and software, is likely to be confronted by strong competitors. Just as the first Brownie was followed by a greater differentiation of camera models for mass amateur photography markets, it is likely that there will be a further proliferation of imaging devices with more sophisticated combinations of sensors, processing power, networking and software. Finally, it is not yet clear what the implications of this emerging Universe of imaging will have for individual and collective senses of identity, place and memory.

ACKNOWLEDGMENTS

Thanks to Amit Kelkar for his role in developing this chapter.

NOTES

1. Félix Guattari, *Chaosmosis: An Ethico-Aesthetic Paradigm*, trans. Paul Bains and Julian Pefanis (Sydney: Power Publications, 1995).
2. Ville Heiskanen. "Apple's iPhone an ad legend before its time; U.S. SALES START TODAY." *National Post*. United States, 29 June 2007.
3. Guattari op cit., 34–35.
4. Guattari op cit., 41.

5. Feature phones are mobile phones that use firmware to control a specific array of features rather than having a full operating system like smart phones. See e.g. Ricardo Bilton and Gloria Sin, "Smartphone vs. feature phone arms race heats up; which did you buy?" ZDNet: Online magazine, *ZDNet*, n.d., http://www.zdnet.com/blog/gadgetreviews/smartphone-vs-feature-phone-arms-race-heats-up-which-did-you-buy/6836

6. William Mitchell, *The Reconfigured Eye: Digital Images and Photographic Truth* (Cambridge, MA: MIT Press, 1992).

7. Resse Jenkins, "Technology and the market: George Eastman and the origins of mass amateur photography." *Technology and Culture* 16 (9 December 1975): 1–19.

8. Ibid., 3.

9. Ibid., 13.

10. Bruno Latour, "Technology is society made durable," in John Law, ed., *A Sociology of Monsters. Essays on Power, Technology and Domination* (London: Routledge, 1991): 103–131.

11. Ibid., 116.

12. Richard Chalfen, *Snapshot Versions of Life*, 1st ed. (Madison, WI: Popular Press, 1987).

13. Nancy Martha West, *Kodak and the Lens of Nostalgia* (Charlottesville, VA: University of Virginia Press, 2000).

14. Peter Buse, "Photography degree zero: cultural history of the Polaroid image," *New Formations* 1 January 2007: 29–44.

15. See Mary Bergstein, *Mirrors of Memory: Freud, Photography, and the History of Art* (Cornell: Cornell University Press, 2010).

16. Walter Benjamin, *Illuminations* (London: Fontana Press, 1992).

17. Roland Barthes, *Camera Lucida: Reflections on Photography* (London: Vintage, 2000).

18. See Michael Naas, "'Now smile': Recent Developments in Jacques Derrida's Work on Photography," *South Atlantic Quarterly* 110(1), 1 January 2011: 205–222.

19. Bergstein, op cit.

20. Jochen Runde, Matthew Jones, Kamal Munir, and Lynne Nikolychuk, "On technological objects and the adoption of technological product innovations: rules, routines and the transition from analogue photography to digital imaging," *Cambridge Journal of Economics* 33(1), 1 January 2009: 1–24. doi:10.1093/cje/ben023.

21. Corey Dzenko, "Analog to digital: the indexical function of photographic images," *Afterimage*, 10 January 2009.

22. Walter Benjamin famously argued (Benjamin op cit.) that mechanical reproduction reduced the aura of the work of art. It seems photography retained something of an aura, which is compromises further with digitality.

23. Kamal A. Munir and Nelson Phillips, "The birth of the 'Kodak Moment': institutional entrepreneurship and the adoption of new technologies," *Organization Studies* 26(11), 1 November 2005: 1665–1687. doi:10.1177/0170840605056395.

24. Rudoph Peritz, "The Roberts Court after two years: antitrust, intellectual property rights, and competition policy," *Criterio Jurídico* 8 (10 April 2008): 283–303.

25. Anon, "Camera-ready Nokia 7650 takes multimedia services to new level," *South China Morning Post* 19 July 2002, 10 (accessed 25 June 2011).

26. Ben Smalley, "Phone cameras lose their virtue," *The Times*, 1 October 2002 (accessed Factiva.com, March 2011), 16.

27. Anon, "Say no to mobile phones with an integrated digital camera?" *Beijing Review* 45(50), 2002: 20.
28. Ibid.
29. Nokia. *Guide to moblogging.* Press backgrounder. 11 February 2005. Online: http://www.nokia.com/BaseProject/Sites/NOKIA_MAIN_18022/CDA/ApplicationTemplates/About_Nokia/Content/_Static_Files/moblog-backgrounder.pdf (accessed 25 June 2011).
30. Barbara Scifo, "The domestication of camera-phone and MMS communication. The early experience of young Italians," in Nyiri Kystof, ed., *A Sense of Place. The Global and the Local in Mobile Communication* (Berlin: Passagen, 2005), 363–374.
31. Kindberg, Ted, Mirjiana Spasojevic, Rowanne Fleck, and Abigail Sellen, "The ubiquitous camera: an in-depth study of camera phone use," *IEEE Pervasive Computing* 4(2), 2005: 42–50.
32. Brendon Slattery, "App overload: Apple passes 300k apps—iPod/iPhone—Macworld UK," Magazine website, *MacWorld UK*, 19 October 2010, http://www.macworld.co.uk/ipod-itunes/news/index.cfm?newsid=3244681 (accessed 27 June 2011).
33. Flickr. "Flickr: camera finder," *Most Popular Cameras in the Flickr Community*, http://www.flickr.com/cameras/ (accessed 23 June 2011).
34. Ben Camm-Jones, "iOS app downloads 'to beat music downloads by March'—PCWorld," Magazine website, *PC World*, 19 January 2011, http://www.pcworld.com/article/216862/ios_app_downloads_to_beat_music_downloads_by_march.html (accessed 27 June 2011).
35. John Duffy, "I spy with my little iPhone . . . ; photographer Matthew Lester pulls latest exhibition out of his back pocket," *Lancaster New Era/Intelligencer Journal/Sunday News*, 23 May 2010.
36. *Hipstamatic* field guide. Online: http://wiki.hipstamatic.com/ (accessed 24 June 2011).
37. Rita Leistner, "Shooting on an iPhone in Afghanistan" in PRI's The World. Online: http://www.theworld.org/2011/05/shooting-on-an-iphone-in-afghanistan/ (accessed 24 June 2011).
38. Gunther R. Kress and Theo Van Leeuwen, *Reading Images: The Grammar of Visual Design,* (London/New York: Routledge, 2006).
39. Ibid., 167.
40. Ibid., 171.
41. Jennifer Preston and Brian Stelter, "Phone cameras credited with helping the world see protests in Middle East," *The New York Times*, 18 February 2011, sec. World/Middle East, http://www.nytimes.com/2011/02/19/world/middleeast/19video.html?_r=2&ref=brianstelter
42. There are a number of iPhone document "scanners" available, including Jot-Not Scanner Pro; Scanner Pro; Turboscan Light; DocScanner; Perfect OCR; Pocket Scanner and Page Scanner, some of which feature optical character recognition (OCR), following the trajectory from image to information by converting text photo into searchable text.
43. Guattari op cit., 37.
44. Tanzina Vega, "Google Goggles app brings users to ad features," *The New York Times*, 15 November 2010, sec. Business Day/Media & Advertising, http://www.nytimes.com/2010/11/16/business/media/16adco.html (accessed 27 June 2011).
45. Jeremy Laurence, "Privacy implications have Google running scared—news, gadgets & tech—the independent," *The Independent*, 14 December 2009, http://www.independent.co.uk/life-style/gadgets-and-tech/news/

privacy-implications-have-google-running-scared-1839884.html (accessed 11 June 2011).
46. Anon, "Facebook: we fumbled face recognition roll-out," *Information-week—Online*, 11 June 2011.
47. Guattari op cit., 51.
48. Guattari ibid.
49. J. David Bolter and Richard A. Grusin, *Remediation: Understanding New Media* (Cambridge, MA: MIT Press, 1999).

8   A Logic of Layers
Indexicality of iPhone Navigation in Augmented Reality

Nanna Verhoeff

INTRODUCTION

In this chapter, I discuss iPhone navigation as a performative practice in mobile and interactive augmented reality tours. I take the iPhone as a theoretical object and examine its specificity as the prime—yet not exclusive—example of today's generation of smartphones, their layered interfaces and the navigation practices enabled by the many applications developed for them. My approach to the trend of digital cartography and augmented reality applications for mobile devices is framed in terms of indexicality. This semiotic concept as defined by Charles Sanders Peirce refers to the intersection of time, place and subject in their relative relationality.[1] In her introduction to a special issue of *Differences* devoted to indexicality and media Mary Ann Doane comes up with the obvious, yet strangely overlooked distinction between the index that comes to us from the past—a trace of things long gone—and the index in the present.[2] This distinction is mobilized and its two temporalities brought in touch with each other whenever the experience in the present is activated by means of reminiscences—a presence of the past *in* the present.

Hence, one kind of index is the *trace*—the real relationship between the present sign and the past, of which it is an imprint. Another one is *deixis*. This linguistic term refers to the here-and-now of the utterance, the situation of discourse (in any medium), which positions a subject, or deictic center. I propose to add to these two a third kind that is brought about by the machine aspect of interface- and software-based indexicality in navigation. Possibility and future-oriented indexicality forms what I call the *destination* index. These three kinds of index can also be considered aspects of a single, complex indexicality, wherein each occurrence of an index can predominantly serve any of the three possible aspects. The iPhone as a mobile and hybrid interface allows for a connection between the here and now, its traces in the past, and a future toward which the subject moves—a connection that evolves in navigation. As such, iPhone navigation involves different modalities of indexicality and a layered temporality in augmented reality.

MAPS AND ARCHIVES

Augmented reality browsing makes use of a combination of built-in features of the iPhone hardware: GPS sensor to locate the position, the camera feed to display the surrounding environment and an internet connection to import geotagged content from an online database. Web-based mobile websites as well as so-called *native* apps—applications specifically designed for the iPhone and effectively distributed through Apple's iTunes App Store—that make use of the technological specifications of the iPhone have opened up a wide range of interactive navigation, touring, and mapping practices. Elsewhere, I have argued that interactive digital maps are part of a visual regime of navigation that goes as far back as early cinema and perhaps even before the cinematic apparatus.[3] Here, I focus on the specific practices of augmented reality navigation particular to the iPhone and the apps developed for it, and in relation to indexicality. The development of locative media enabled by handheld mobile screens marks a shift toward indexical, deictic navigation. Moreover, the indexicality of navigation, I argue, involves a changing status of the trace.

The basic principle of screen-based navigation is that we see how we move—while how we move enables this vision. As such, in iPhone navigation both space and time unfold in practice. Its performative cartography is a process of simultaneous image capture and experience. This mutually constitutive relationship between seeing and moving—between perception and doing—is a fundamental principle in real-time, digital cartography. It is in movement that the map takes shape. Reading space thus requires navigation, rather than the other way around.

With touchscreen, camera, accelerometer, compass, GPS, network connectivity and the diverse mapping applications that are currently being developed, the iPhone essentially has become a hybrid cartographic interface for navigation.[4] This navigational model brings about a shift in cartography as we have come to know it. Originating in the art of making maps, but putting forward a new regime of understanding and representing space, mobile cartography has infused spatial representation with a distinct temporal and procedural dimension. The iPhone testifies to the advent of *performative cartography*, as it offers a dynamic map that emerges and changes during its user's journey. Both categories of physical and information space are inextricably connected in a hybrid *screenspace*, as I call it, of this new form of mapping. Producing images while viewing them, the user-navigator engages physically with the iPhone in a temporally dynamic and spatially layered process. This is a cartographic collapse of making-while-navigating maps. Because this process transforms the viewer to actual user, it exerts performativity. Because users *must* act in order to use the map, they become, indeed, performers. Thus, the performativity of interactive digital cartography results from the actual performance of navigation.[5]

This performative process engages traces from the past in the present of locative presence. The ancient topos of memory as a place is presently revived in the form of digitalized archives available for on-site navigation. Contemporary cultural archives and museums are now exploring the possibilities of locative media and mobile platforms and offer content from their collections in geotours or augmented reality applications. iPhone apps that make use of museum collections or other visual archives are the 2010 *Museum of London: StreetMuseum* iPhone app, or the *Oorlogsmonumenten in Beeld* app developed by the Dutch *Erfgoed in Beeld* (Imaging Heritage) project for accessing information about war monuments in the Netherlands. The Netherlands Institute for Architecture has also launched *UAR*, the Urban Augmented Reality application, which I will discuss in more detail below. Also, other museums with non-locative collections seek possibilities for interactive apps to provide access to their collections, such as the American Museum of Natural History's app *Dinosaurs: The American Museum of Natural History Collections*. The Stedelijk Museum in Amsterdam has invested in a large project, ARtours, which involves not only offering access, but also developing contributions to their collection by commissioning digital art for augmented reality platforms.[6]

Aside from these institutional archives and museums that experiment with innovative interfaces for their digital collections, think also of more personal (bottom-up) or collaborative online databases with travel itineraries, photos and personal annotations and applications that are used to share this material. Digital archives need interfaces to become "performable"—that is, to be made available for performance and participation. This potential performativity and the possibility for contributions to archives in the form of user-generated content, is collaborative, interactive and participatory. In these projects, interfaces are made mobile so that memory becomes attached to places again, revamping the ancient topos with an update: from memory as a place, memory becomes a machine.[7] These "practical" interfaces are perhaps a subset of what Manovich has called more broadly "cultural interfaces"—interfaces that construct a model of the world.[8] At the same time, the incipient status and great diversity, as well as rapid changes in technology, also make the practical interfaces to our digital collections relatively unstable and temporary.

This relationship to the past—our cultural archive—in screens of navigation as producers of traces shows us that in our present visual culture where viewing and making collapse, past (trace) and present (deixis) also become entangled. Hence, the notion of the cartographic becomes double. It points to the property of screen-based moving-image media, as well as the characteristic of our creative engagement with these screens in cartography as practice. In what I call screenspace, we are simultaneously narrator, focalizer, spectator, player, and when all these functions join forces, most fundamentally, navigator. Amidst the fast-changing gadgets, platforms and applications for mobility and navigation, we can discern a new location-based, cartographic archival logic—a logic of layers.

THE iPHONE INTERFACE

As a mobile and hybrid device, the iPhone's interface serves to make this logic operable. It mediates time and space according to indexicality. For this, it features a complex and layered structure of characteristics and affordances, which allows for a broad range of interactive practices. Thus, the iPhone first of all may prompt questions about its specificity as a *hybrid object*.[9] It is this hybridity that engages the user in ways that enhance indexicality. Hybridity relates to iPhone *interfacing*, an entanglement of technologies, applications and interactive practices. This interfacing takes place within what one might call a mobile screening arrangement, or *dispositif*—a term derived from early French film theory, developed by Jean-Louis Baudry to provide a theoretical construct of what is often called in English language the cinematic apparatus, and helps us to analyze the material and spatial specificity of the "set-up" within which screens operate.[10]

The iPhone as a hybrid object is embedded within a mobile arrangement that encompasses both the perceptual positioning of the (mobile) user and the physical (interactive) interfacing with the screen. The screening arrangement in motion, taking place within public space and making connections with this space, establishes a mobile sphere: a private/public space that is marked by (individual) mobility and (networked) connectivity, and which is constructed within the (mobile) arrangement of user, location, and device. Given the use of the iPhone for navigation both with and on the (touch) screen, the mobility of the device makes it both a haptic and visceral interface: the entire body of the user is incorporated in mobility and space-making.

This puts the user at the center of a deictic network. The iPhone is a cartographic interface for the simultaneous navigation of both on-screen and off-screen space. In a marked difference to historical cinematic and televisual screens the iPhone enables a navigation of both the machine itself and the physical space surrounding its user. It encapsulates the user and the machine within a mobile dispositif of navigation. Hence, it positions the user within a mobile sphere implying an ambulant locatedness and, consequently, flexible site specificity. Popular iPhone applications like *Foursquare*, where one can check in on specific locations, or *TweepsAround*, which uses an augmented reality interface for (on-screen) visualizing of the (off-screen) presence of Twitter connections in the area, testifies to an interest in mobile location-marking based on indexicality. To clarify my terminology here, in analyzing the (layered) interface, the applications and practices of navigation, I will use "hybridity" for the iPhone as device, a "layering" of its levels as interface, and the term "augmented reality" for both the specific genre of applications I focus on, as well as the particular spatiality of the screenspace that is navigated with the iPhone.[11]

The development to an obvious and keenly felt bodily navigation such as is evident in augmented reality mobile applications foregrounds how the indexicality is deployed in hitherto unseen complexity. The iPhone's mobility and physicality, I argue, point toward the performative and embodied features of interactivity as being characteristic of navigation generally. From this point of view, navigation not only entails a spatial decoding of map information, orientation, and mobility, but also a cultural trope which makes our sense of (spatial as well as temporal) *presence* centrally deictic. Hence, screenspace is activated by the simultaneous construction of on-screen and off-screen spaces when traversing in fluid motions with navigation devices in the user's hands.

This interface for navigation is layered. While intricately connected and hard to separate and isolate, conceptually there are three (non-hierarchical) levels, all essential for navigation. First, navigation comprises the *internal interfacing* of applications: the back-end operating system and software, and consequently, the *processing* of data. The Google Maps application programming interface (API), for example, is suitable for mapping applications, because it provides tools for web-application hybrids, or *mashups*. These allow the integration of data from different sources within, in this case, the mapping environment of Google Maps.

The second layer of the interface concerns the spatial positioning and connectivity of the apparatus in relation to physical as well as data space: the interface of the internal instruments of the iPhone that connect it to external space. This level of the interface communicates between the hardware of the device and its surrounding "reality." It includes an *inertial* navigation system, which according to Oliver J. Woodman is "a self-contained navigation technique in which measurements provided by accelerometers and gyroscopes are used to track the position and orientation of an object relative to a known starting point, orientation and velocity."[12] This ability to calculate position and orientation is necessary for, for example, gravimetric (rather than marker-based) augmented reality applications as interfaces for location-based data, or *ambient intelligence*. Moreover, internet connectivity also positions the device via wireless connection. The second layer of the interface, then, concerns *connecting* and *positioning* the interface whether based on inertial, absolute, camera-based or wireless technologies.

Positioning is communicated to the user who may see the on-screen image tilt, or who may find a representation of her position and movement signified by an "arrow" in the on-screen maps, and then may read this orientation and act or move accordingly. This is all taking place on a third level of the interface I call *user interaction*, enabling the communication between the user and the internal operation of the device (first level) as it is connected to the space surrounding it (second level). It is on this level that indexicality becomes, so to speak, embodied. While the first level of the applications interface also includes the software operation of the graphical user interface (GUI), the way in which this data is visualized and made understandable operates at this third level of user interaction. In the case of navigation this level entails the way spatial information is represented on the screen and interacted with by the user. Think, for example, of the ways roads or larger maps are represented on screen in navigation applications for the iPhone. Then, one can adjust, move or zoom in or out by using swipes, taps or pinching movements with fingers on the multi-touch screen.

Significant for the touchscreen of the iPhone is that at the level of user interaction, it is an instrument for both input and output. This is the level of access to and the experience of data; the action literally takes place on the screen. The dynamic horizontal or vertical scrolling of screen content establishes a connection between the image on-screen and its off-screen spaces: the frame is always a detail from a larger whole, and the map is always larger than the part or detail that is displayed on screen. Objects can be moved outside and brought into the frame by the swipe of a fingertip.

Seen within the layered constellation of its interface, understanding the iPhone thus requires a triple perspective: it is a machine that processes and combines data; a sensor that connects and positions data; and a medium that produces perception. In semiotic terms, working the interface is an essentially indexical operation. While walking and using the iPhone for interactive tours, the different layers of the interface operate together: location-based information is processed and communicated to the user via the screen. The integration of these processes (data processing, spatial positioning and connectivity, and the communication with the user) is the *condition of possibility* for navigation as performative cartography. Understanding the indexical as trace, deixis and destination working together helps to grasp what is new, what is a further development of earlier processes, and how people make use of such devices. In what follows, I investigate how iPhone navigation constructs a performative archival space in which temporal layers intersect.

DESTINATION INDEX

When we consider earlier debates about photographic media technologies, we can see how indexicality has been investigated in terms of authenticity or veracity of the image in the time of mechanical reproduction. The photo-chemical base of the photographic and filmic image is understood as proof of a prior existence. Doane traces this discussion in light of the "crisis of legitimation" of digital media. This existential aspect of the index, as trace, has been foregrounded in the ontological conception of Pierce's positioning of indexicality in semiotic logic.[13]

Doane, however, brings together the two very different characteristics of the index that we can discern in Pierce's writing on the index, that is, the temporality of the index as trace and deictic directionality. She problematizes the issue of authenticity by proposing a dialectic of these two sides to indexicality: the implied temporality of the index as imprint (what Barthes calls the "this-has-been" of photography) and as indicator: "look here." This indication has a very forceful presence, if not present. In Peirce's own words:

> [T]he sign signifies its object solely by virtue of being really connected with it. Of this nature are all natural signs and physical symptoms. I call such a sign an *index*, a pointing finger being the type of the class. The index asserts nothing; it only says "There!" It takes hold of our eyes, as it were, and forcibly directs them to a particular object, and there it stops. Demonstrative and relative pronouns are nearly pure indices, because they denote things without describing them.[14]

The interplay of deictic directionality and the meaning of the trace are very clear in the case of digital navigation and can be clarified through geotagging. The principle of geotagging is a key innovation in digital cartography and the development of mobile technologies and applications. The attachment of information to objects and locations, and vice versa, based on global positioning (GPS) makes a whole range of uses possible: from adding locations to photographs, to providing on-site information and adding hyperlinks to maps. Geo-coordinates make it possible to annotate location-specific information to objects, or to connect archives with a vast information database to particular locations. The digital may have shaken the grounds of this trust in photographic authenticity yet it has also opened up the trace to include not only a link with a moment of prior existence, but also the virtual presence of possible destinations.

On the iPhone, geotags can activate different spatial and temporal layers. Dots on the screen/map unfold, as spatio-temporal hyperlinks. The city becomes a navigable and clickable screenspace, a terrain of pop-ups that are triggered by real-life avatars in the physical world whose movements are traced on-screen by GPS. Two-dimensional maps are a flat and still representation of space within a fixed frame, based on a fixed scale, with an abstract perspective. The digital map is dynamic, layered, expandable, mutable, and with flexible points of view. Geotags bring together all levels of the hybrid interface of the iPhone. By combining and locating data, geotags are visible mainly through their effects in screen navigation, activating content by directional and proximal physical movement in the world. Marc Tuters and Kazys Varnelis speak of two kinds of cartography in the broader genre of locative media: annotative cartography, based on tagging, and phenomenological cartography, based on the tracing of movement. This is close to my terminology here, although I wish to analyze the merging of these two forms in performative cartography, as it is made possible by the hybridity of interface of the iPhone.[15]

Geotagging activates the old principle of the trace, well known from photography. Geotagging photographs entails adding GPS coordinates of the place the picture was taken to a data file in which the digital image is stored. Including a time stamp in the photographic data underscores the temporal as well as a geographical aspect of tagging. It allows for a mnemonic mobility by placing and tracing of digital footprints. We can understand this implication of memory as reinstating the "lost" indexicality of photography. Once the (analogue) photograph was a literal imprint of light, which allegedly proved spatio-temporal reality and thus provided the image with "authenticity." Digital photography lost this direct relationship between reality to image. However, today we can restore this rupture by attaching coordinates in time and space as digital information. This location is not necessarily close to what is portrayed in the photograph, to the object of the image. But it does locate the object as well as the camera and photographer in reality. These coordinates constitute the image's digital footprint—its trace.

Geotags make it possible to retrace these digital footprints. In their capacity to create locative and semiotic connections, tagging entails a potential for participatory engagement. People can make their own personal archives, use them for exchange, or participate in creating collective archives. Tagged "mobile mementos" make collective image gathering possible, based on the collection, connection, or contribution of information derived from large, social databases. This is where the trace (of the past) joins the deixis (in the present), but also points to a future. The tagged images of past travels become destinations in future navigation. When we consider the geotagged photo as destination for navigation, the image shifts in temporal-indexical status. The geo information actually refers to the location of the device at the moment of taking the picture. It does not point to a presence of the object in the image. The image becomes a trace for the point of view, pointing to a presence behind the camera: the subject who looks. The geo-coordinates take the traveler to the same location, making it possible to step into the footsteps of the deictic center. This is how the trace, in shifting to destination, becomes deictic.

DEIXIS OF NAVIGATION

Navigation as way-finding entails constantly registering presence (where am I?). But rather than focusing on the trace of the past, navigation is geared toward deciding where to go next. Hence destination (where will I go?) becomes the new center of indexicality. Space is constructed in this indexical reading of space where the three temporalities merge.

Let us look at the consequences of the fact that the present moment is indexicality signified. Deixis is a term borrowed from linguistics to explain how language is context-dependent. The focus on deixis has shifted the conceptions of language dramatically. In fact, as Émile Benveniste has argued, deixis, not reference, is the essence of language.[16] Deictic words, or shifters, function as mobile focal points, often within an oppositional structure such as "here," implicitly opposed to "there." Deixis indicates the relative meaning of the utterance, tied to situation of utterance, an *I* in the *here* and *now*. They have no fixed, referential meaning. Deixis establishes the point of origin, or deictic center, of the utterance: the I who speaks, as well as its point of arrival, the *you* that is spoken to. These terms are by definition mutually exchangeable. Moreover, or consequently, deixis frames the statement in temporal (*now*) and spatial (*here*) terms.

As such, deixis helps set up the world to which the text relates. In contrast to nouns or adjectives, deictic words, or shifters, only have meaning in relation to the situation of utterance. Their meaning is produced through indication rather than reference—think of pointing. Personal pronouns of the first and second person—I, we or you—are shifters. But *he, she* or *it* are not. The latter, although also in need of identities to fill them in, do not change when the situation of utterance changes. But when *I* speak and *you* answer, *you* become *I*, and *I, you*. Hence, if we do not know who is speaking, the first and second person pronouns have no meaning. Similarly, we cannot *place* the meaning of such words as "over there" or "right here" if we don't know from where the speaker is speaking. Nor can we *time* the meaning of "yesterday" without a determined temporal frame.

I deploy these examples of shifters to suggest that the intersection of time, place and subject is their primary anchor. Therefore, while the term was first introduced in linguistics, the perspective on the construction of space, time and subjectivity in screen-based spectatorship is particularly useful for analyzing how the spectator or user—here, navigator—is bound to the screen image. At the heart of deixis lies the interaction between an "I" and a "you" whose positions are essentially exchangeable. This can be applied to user-screen interaction and sheds light on the particular performativity involved in navigation—a reciprocal performativity of making space in a (subjective) engagement with it and the constituting subjectivity in the reading of space.

Hence, the screen image is not simply presented as from an internal point of view—a diegetic perspective—but also produces the subjectivity of the spectator (the "I" doing the looking) as well as of the "I"'s "you," the second person who mutually constitutes and affirms the "I." A screen image is what tells us "about," and thus constitutes, a (fictionalizing) gaze that emerges through the inflection of the vista that invests it with subjectivity. This inflection can also be called focalization, as a term that expresses this mediating and subjectivizing function, a visual equivalent of deixis.

Screen-based deixis concerns the situatedness of the image in the present of its emergence. In the case of the iPhone, this is can be a simultaneous capture and transmission by the live feed from the phone's camera in augmented reality browsers such as *Layar, Wikitude, Junaio* or *Acrossair*, for example. Or it can be the basis of geotagging the temporal and spatial position of the device at the moment of taking a picture. This tracing of deixis reverses the origin perspective so engrained in an ontological understanding of indexicality. Because pointing to the *here* of the object of the photograph positions the device/user *in relation* to the image, this *here* constitutes the presence (and present) of the device as an indexical interface.

In terms of the two aspects of indexicality of trace and deixis, in these situations the trace implies deictic positioning, shifting from the trace as an origin of perspective to a present and possible future of perception. Geotagging traces the position of a viewing position rather than the vista itself and provides the coordinates for future navigation. As such, it is a typical practice of deixis, no less than analog photography has been established as the key practice of the trace-index. It is thanks to its deictic indexicality that geo-tagging allows the trace of the past (the moment of capture) to become the deictic index in the present. It is from the present position that the traveller looks towards a third temporality of the index, pointing toward the future destination of the traveler. Indexicality pointing to a potential future position is what I call the destination index.

The close connection between screens and maps and the pertinence of deixis in a performative conception of cartography becomes clear in Tom Conley's discussion of the analogy between the cinema screen and cartography as "locational imaginings."[17] Conley points out via David Buisseret's use of this term that cartographic media locate subjects within the places they represent.[18] Conley then brings this in relation to cartographic deixis, after the French semiotician Christian Jacob, whose seminal book *The Sovereign Map* Conley translated from the French.[19] Jacob bases his theory of the deictic nature of maps on Benveniste's linguistic theory. The linguist proposed this term to account for the implication of the speaker in what is being said. The point of bringing in the terminology of deixis is that deictic meaning cannot be understood without taking into account the situation of utterance or the image itself. This leads to the key phrase "you are here" that defines the cartographic act.

A quintessential deictic marker in digital cartography is the arrow-avatar, the digital "you are here" that we know from analogue maps, positioning information in context-specific references. This positioning goes both ways: it calibrates the deictic center (*you* are *here*) as well as the relative positioning of the information the *I* can find on the screen/map. Note that this is sometimes complicated by the "we are here" in some maps that point to possible destinations, like ads providing information about the location of a certain restaurant or shop. The *we* is not inclusive in the present, but inclusive in the possible future present: *we* are *here*; if *you* come *here*, you can become part of *we*.

The map requires that the subject decodes the (imaginary) phrase *you are here* into *I am here*. The map is only usable once the subject knows where the *I* exactly is positioned. The act establishing a deictic center is at the heart of the navigation system. I contend that performativity—in this context, but more generally speaking as well—requires an activation of deixis: positioning a deictic center within a visual, spatial field. Interactive navigation with the iPhone as interface/map visualizes this situation in two ways. The interactive map embodies the user's position as focalizer of the map. It also reflects back what the user does with the map, what itinerary the user creates and simultaneously travels.

A personal anecdote of the opposite—in which deixis failed through a lack of training in mobile spatial perception and hence, a lack of performative engagement—is my very first driving lesson. It was my impression that everything I saw through the car window was so much like a television image, that it did not seem to be really present: out there and right now. What I saw was a screen image, a representation and not really there. Precisely the fact that I was in the driver's seat, literally and figuratively, triggered this visual experience. I had been in cars before, obviously, but this visual experience was new. I did not know how to deal with what I saw through the window, how to act, or how to see myself in relation to the framed image. I was unable to think deictically. This lack of deixis, or deferred deixis, characteristic of a more traditional regime of representation, put me as *I* outside of the image. The world in that regime is, literally, objective—a reality that is impossible to touch.

This example makes clear how a performative visual regime of navigation, requiring deictic activation, establishes a physical engagement that makes the experience of looking a form of haptic engagement. This engagement brings together the aspects of agency—the doing—and the experiential—the seeing and feeling. It is the haptic engagement, understood as form of interactivity and as experience, which is significant for mobile screen gadgets. Haptic experience takes place at the intersection of touch and physical interaction of the experience of the device, on the one hand, and the agency in and experience of spatial unfolding on the other. It is in haptic engagement that the creative meets the cartographic. This haptic relationship to space, then, is a particular quality of iPhone navigation that involves more than just its touch screen but puts the whole layered interface at work—including the subject that uses the interface.[20]

Augmented reality—perhaps a safer environment for failed deixis than the view through a windshield while behind the wheel—also provides a framed image, and opens up the possibility of viewing objects that are not physically present. In displaying visual data, however, augmented reality applications do establish presence, and, hence, subjectivity.

A LOGIC OF LAYERS

Augmented reality applications for the iPhone exemplify the way in which the hybrid, layered interface of the device can be used to visualize and access geo-specific content. Augmented reality browsers such as *Layar, Wikitude, Junaio* or *Acrossair* are rapidly expanding the possibilities of (consumer) augmented reality. They offer browser applications on smartphones that have a video camera, GPS, a compass and an orientation sensor, entailing new ways of engaging with screenspace by effacing the map representation and using direct camera feed with a superimposed layer of data. Augmented reality browsers make it possible to browse for data directly within "reality" as it is represented on the screen. The camera eye on the device registers (rather than captures, or imprinting as a trace) physical objects on location, and transmits these images in real time on the screen, where the image is combined with different layers of data in image or text. Information is, thus, superimposed on a real-time image on screen.

For example, the *UAR* (Urban Augmented Reality) app of the Netherlands Architecture Institute (NAi) makes use of the *Layar*-infrastructure for an augmented reality browser, which is connected to their digitized archive. It shows 3D visualizations of buildings on location that were once there in the past, information about current buildings, visualizations of plans for the near future, or designs that were never actually built at all.

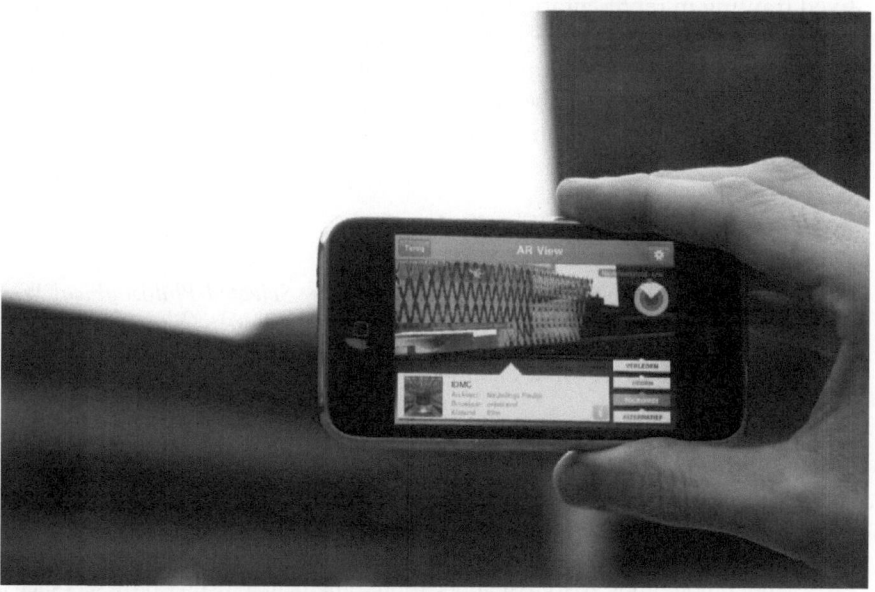

Figure 8.1   Augmented reality view of the future: the IDMC building in The Hague, the Netherlands. Image: Netherlands Architecture Institute (NAI), 2011.

In the hybrid screenspace that this tour constructs by layering information on the live camera feed of the iPhone, the present, past, future and even the past future, coincide:

> With 3D modelling you can see buildings in the heart of the city, where they stood in the past, or will be built in the future. You can walk around them and explore them from all angles. Using UAR, you can see what isn't there! . . . You can review plans, see historic photo material, read biographies of architects and take walking routes through the city.[21]

The hybrid image of augmented reality combines digital data with a live camera feed. The screen is in effect both a transparent window and an opaque screen for data overlay. The camera view enables you to see your surrounding space in relation to and framed by the camera. This layering of the screen image gives a spin to Anne Friedberg's notion of the virtual window—the multiplication of frames (windows) on the computer screen, frames within frames.[22] The layering of data within the frame is perhaps a multiplication of frames that complicates the frame analogy altogether, as the resulting image is a hybrid combination of images rather than insertions of frames within frames.

Above I pointed out how digital navigation can be understood as a cultural trope which makes our sense of presence centrally deictic, determining one's current position, with a forward-slanted orientation towards possible destinations. This trope builds on the logic of layers, breaking with the regime of fixed framing in representation. This logic is demonstrated by iPhone navigation, bringing together trace and deixis in the destination index. Navigating augmented reality gives a variable spatial and temporal texture to space. It transposes the archive to a live stream of performative, spatial experience—an experience of an intersection of past, present and future.

NOTES

1. Charles Sanders Peirce, *The Essential Peirce: Selected Philosophical Writings*, Vol. 1, Nathan Houser and Christian Kloesel, eds. (Bloomington: Indiana University Press, 1992).
2. Mary Ann Doane, "Indexicality: trace and sign: introduction," *Differences* 18(1), 2007: 1–6.
3. Verhoeff, Nanna, 'Screens of Navigation: From Taking a Ride to Making a Ride', *Refractory: A Journal of Entertainment Media*, 12 (special issue on Screenscapes, 2008.
4. These applications are categorized under "navigation" in the iTunes app store, or sometimes "travel" or "entertainment," and range from straightforward way-finding tools, such as *Motion X GPS*, or the *Navigon* or *TomTom* iPhone apps, to tourist applications, augmented reality browsers such as *Wikitude* or *Layar*, or apps for treasure hunts such as *Geocaching* or *Garmin OpenCaching*, and social-networking apps such as the "social travel guide" *Gowalla*, or *Foursquare* and *TweepsAround*.

5. Jeremy W. Crampton locates a performative turn in cartography and understands maps as performative practices rather than as objects. See "Cartography: performative, participatory, political," *Progress in Human Geography* 2009: 840–848. For the distinction between performance and performativity, see "Performance and performativity" in Mieke Bal, *Travelling Concepts in the Humanities: A Rough Guide* (Toronto: Toronto University Press, 2002): 174–212.

6. About the *ARtours* project and more generally about augmented reality and the museum experience, see also Margriet Schavemaker, Hein Wils, Paul Stork, and Ebelien Pondaag, "Augmented reality and the museum experience," *Museums and the Web 2011 Conference,* 6–9 April 2011, Philadelphia, PA, USA http://conference.archimuse.com/mw2011/papers/ augmented_reality_and_the_museum_experience (accessed 3 July 2011), and their website http://www.artours.nl/. Nancy Proctor, Head of Mobile Strategy and Initiatives at the Smithsonian Institution curates a good online collection and "forum for conversations about mobile interpretation—media & technology—for museums and cultural sites" at http://museummobile. info/ (accessed July 2011). Another good resource is the *Museum 2.0* blog by Nina Simon, Executive Director of The Museum of Art & History at the McPherson Center in Santa Cruz, CA, at http://www.museumtwo.blogspot. com (accessed July 2011).

7. A now classic study on the art of memory and historical practices of memory retrieval is Frances A. Yates, *The Art of Memory* (Chicago: University of Chicago Press, 1966).

8. Lev Manovich, *The Language of New Media* (Cambridge, MA: MIT Press, 2001), 63–93. Manovich points out a layering of digitization, or in his conception, a dialogue of software operation and human activity in the operation of these interfaces. His conception of layering is that of a working together of distinct registers (cultural, computer or human). Here, I wish to reserve "layering" for a hybrid composition of object and operation (interface) and of spatial construction of navigation, as product of using the interface for interfacing. His conception of a particular machinic and interactive quality of interfacing is pertinent to intermediality in theater and performance, as pointed out by Freda Chapple and Chiel Kattenbelt, "Key issues in intermediality in theatre and performance," in Freda Chapple and Chiel Kattenbelt, eds., *Intermediality in Theatre and Performance* (Amsterdam: Rodopi, 2006), 11–25.

9. Elsewhere, I have proposed to speak of mobile gadgets as theoretical consoles—rather than theoretical objects. Nanna Verhoeff, "Theoretical consoles: concepts for gadget analysis," *Journal of Visual Culture* 8(3), 2009: 279–298.

10. Jean-Louis Baudry, "Ideological effects of the basic cinematographic apparatus" and "The apparatus: metapsychological approaches to the impression of reality in the cinema," in Philip Rosen, ed., *Narrative, Apparatus, Ideology* (New York: Columbia University Press, 1986), 286–318.

11. Adriana de Souza e Silva argues for the term hybrid space, rather than augmented reality. She focuses on the social consequences of the blurring boundaries between virtuality and physical space—in particular as a consequence of connectivity. She takes up, amongst others, Lev Manovich's approach to augmented space but proposes a different terminology for our engagement with these spaces: "From the merging of mixed reality and augmented spaces, mobility, and sociability arises a *hybrid reality.* It is exactly the mix of social practices that occur simultaneously in digital and in physical spaces, together with mobility, that creates the concept of hybrid reality." Adriana

de Souza e Silva, "From cyber to hybrid: mobile technologies as interfaces of hybrid spaces," *Space and Culture* 9(3), 2006: 261–278); Lev Manovich, "The poetics of augmented space: learning from prada," http://www.noema-lab.org/sections/ideas/ideas_articles/pdf/manovich_augmented_space.pdf. (accessed 3 July 2011). I do agree with her argument for using "hybridity" for the "outcome" of engaging with mobile technologies such as the iPhone.

12. Oliver J. Woodman, "An introduction to inertial navigation," *Technical Report UCAM-CL-TR-696*, (London: University of Cambridge Computer Laboratory, 2007), 4. About ambient intelligence see Emile Aarts, Rick Harwig and Martin Schuurmans, "Ambient intelligence," in Peter Denning, ed., *The Invisible Future: The Seamless Integration of Technology in Everyday Life* (New York: McGraw-Hill, 2001), 235–250.

13. Mary Ann Doane, "Indexicality: trace and sign: introduction," 1.

14. Charles Sandes Peirce, "On the algebra of logic: a contribution to the philosophy of notation," *The American Journal of Mathematics* 7, 1885: 180–202. Reprinted in *The Writings of Charles S. Peirce,* Vol. 5, compiled by the Editors of the Peirce Edition Project (Bloomington: Indiana University Press, 1993), 162–190; Mary Ann Doane, "The indexical and the concept of medium specificity," *Differences: A Journal of Feminist Cultural Studies* 18(1), 2007: 128–152.

15. Marc Tuters and Kazys Varnelis, "Beyond locative media," *Leonardo* 39(4), 2006: 357–363.

16. See Émile Benveniste, *Problems in General Linguistics,* trans. by Mary Elisabeth Meek (Coral Gables: University of Miami Press, 1971). For an excellent overview of the ins and outs of deixis, see Stephen C. Levinson, "Deixis," in Laurence R. Horn, ed., *The Handbook of Pragmatics* (Oxford: Blackwell Publishing, 2004), 97–121. Levinson considers deixis as coextensive with indexicality, which he considers a larger category of contextual dependency and reserves deixis for linguistic aspects of indexicality (97–98).

17. Tom Conley, *Cartographic Cinema* (Minneapolis, MN: University of Minnesota Press, 2007), 2.

18. David Buisseret, *The Mapmaker's Quest: Deciphering New Worlds in Renaissance Europe* (New York: Oxford University Press, 2003).

19. Christian Jacob, *The Sovereign Map: Theoretical Approaches in Cartography Throughout History* (Chicago, IL: University of Chicago Press, 2006).

20. Conley suggests that the making of space in navigation happens in the process of perception and, as such, has a haptic quality. Conley, *Cartographic Cinema*, 20. About the use of mobile touchscreens, Heidi Rae Cooley argues that it transforms the practice of screening as tactile *activity* into a haptic *experience* of this practice. Heidi Rae Cooley, "It's all about the fit: the hand, the mobile screenic device and tactile vision," *Journal of Visual Culture* 3(2), 2004: 133–155.

21. Description of the application *UAR*, from iTunes, http://itunes.apple.com/app/uar/id371459375?mt=8# (accessed 3 July 2011).

22. Anne Friedberg, *The Virtual Window: From Alberti to Microsoft* (Cambridge, MA: MIT Press, 2006).

9 Touching the Screen

A Phenomenology of Mobile Gaming and the iPhone

Ingrid Richardson

INTRODUCTION

> The world inside the phone becomes more vivid and engaging every day. It wants your ears, eyes, thumbs—all of you.[1]

With every change to our technological interfaces, there is a corresponding modification to perceptual reach and communicative possibility. The shift from analogue to digital technologies and forms of media over the past fifty years has mobilized a critical transition in how we relate to and make meaning of the world. One of the most significant cultural effects of this translation of image and information into digital code is the increasing predominance of telepresent screen interfaces and media forms. As Jon Olav Eikenes and Andrew Morrison observe, contemporary screen technologies are dynamic and multi-modal, literally "sites" for a range of activities and situated uses.[2] Today, mobile media devices and "wearable" screens are becoming both increasingly ubiquitous and personalized, penetrating and transforming everyday cultural practices and spaces, and further disrupting distinctions between private and public, place and space, ready-to-hand and telepresent interaction, actual and virtual environments. In this chapter I examine how one such technology—the touchscreen smartphone exemplified by the iPhone—has had an impact on our embodied perception of space, place and (tele)presence. In particular, I consider the spatial and locative effects of pervasive and networked mobile gaming—an occasional and high-end practice—and then offer a comparative analysis of the much more mundane and common activity of casual mobile gaming.

Upon the initial release of the iPhone in 2007, Joel Mace and Michael West identified three major differences that distinguished it from competing mobile phone products: the large capacitive touchscreen and keyboard, integration into the already existing iTunes store (as a "high-end" model of the iPod with phone capability) and the inclusion within the device of Apple's web browser, Safari.[3] In 2008, Apple also released the software development kit (SDK), soon to be followed by the opening of an online App Store one day before the iPhone 3G became available.[4] As

West and Mace document: "In the first six months, the store attracted more than 15,000 applications and 500 million downloads, and three months later (April 2009) those figures had doubled to 30,000 and 1 billion, respectively; in November 2009, the figures reached 100,000 and 2 billion"—the most popular were customized interfaces for existing websites (eBay, MySpace), locative web-enabled services and standalone applications (especially games).[5]

It is this combination of factors—touchscreen, App Store and web browser—that heralded a new type of mobile phone experience: the adaptation of a mobile device to the internet, with an application marketplace that brought with it the affordances of networked computing.[6] As West and Mace point out in their analysis of the iPhone's market success, the increase in mobile internet use was a result of an important shift in thinking: from the provision of an internet tailored to mobile screens (e.g., evidenced by micro-browsers and provider portals) to the provision of the "real internet" on the mobile device, initiated with the iPhone. They state:

> Three weeks before the release of the first iPhone, Steve Jobs predicted, "people want the real Internet on their phone" . . . Jobs was eventually proven correct: when given web browsing that was substantially similar to the browsing experience on a PC, mobile web usage went up dramatically. The success of the iPhone demonstrated that what was holding back demand for mobile data services in the US was not the creation of new mobile-specific value networks, but the delivery of devices and networks that were capable of delivering a convincing approximation of the familiar wired Internet.[7]

Prior to the iPhone, providers and handset makers sought to create a tailored internet—via provider-based portals or "walled gardens"—specifically designed for mobile devices, under the assumption that the large screen experience of internet computing and browsing would not translate effectively onto small mobile screens without keyboards.[8] Apple's strategy focused instead on a reversal of this assumption, by developing a device adapted to the existing wired web; as West and Mace argue, "The killer app for the mobile Internet was the same as for the wired Internet: a web browser" that could deliver a "convincing approximation of the familiar wired Internet."[9] By late 2007, the iPhone became the most common mobile browser on Google; two months later, according to Google, "the iPhone had fifty times as many Internet searches as any other mobile handset."[10]

As a number of theorists of mobile culture have noted, the iPhone has worked to escalate the popularity of location-based services and games, and thus the merging of online and offline, physical and digital, co-presence and telepresence. Moreover, it has in many ways set the standard as the mobile exemplar of the *ludic turn* in contemporary culture, a conduit and container of numerous playful and often user-generated applications.[11] In what follows I consider how the locative and web capability of the iPhone, the proliferation of apps in combination with touchscreen functionality, has facilitated two modalities of gameplay—location-based and casual—and how this has had an impact on the experience, perception and embodiment of mobile gaming.

A PHENOMENOLOGICAL APPROACH

The analysis to follow is fundamentally informed by a phenomenological approach. In Maurice Merleau-Ponty's perceptual and what I would call *artifactual* epistemology, the corporeal schema, or our lived experience of perceptual reach and bodily boundaries, is always already extendible through artifacts and technologies.[12] It is the corporeal schema that accounts for the body's capacity to intertwine with the world, to integrate, internalize or *intercorporealize* seemingly external objects, spaces and environments into our corporeal activities. This coupling of tools and bodies is effectively articulated by the term intercorporeality, which describes the irreducible relation between technologies, embodiment, knowledge and perception. Our corporeal schema is not determined by the boundaries of the material body but rather reflects the way that our corporeality extends and withdraws—changing its very reach and shape—in its dynamic apprehension of tools and things in the world. Merleau-Ponty argued that this schematic is inherently open, allowing us to incorporate technologies and equipment into our own perceptual and corporeal organization. As Merleau-Ponty famously claimed, the body "applies itself to space like a hand to an instrument,"[13] an application that depends as much on the specificities of perception and bodily movement as it does on the materiality of the tool in use. It is our somatic *openness* to the "stuff" of our environment that allows us to incorporate technologies and equipment into our own corporeal organization.

The human-technology relation is thus irreducible—being-in-the-world is not containable in either term but only explicable as a *relational ontology*; the condition of our existence is as socio-technical or techno-corporeal hybrids. In these terms, we can see how the cultural and technological specificity of media interfaces and apparatuses can be understood as deeply integral to our individual and collectively realized corporeal schemata. Indeed, considering the number of hours that many people spend engaging with media in contemporary life, the body-screen relation in particular may be one of our most significant human-technology relations. This chapter is concerned with the human-technology relation specific to bodies and mobile media, and how mobile devices "dilate" the body into network and game space. As George Simmel observes: "A person does not end with the limits of his physical body or with the area to which his physical activity is immediately confined but embraces, rather, the totality of meaningful effects which emanates from him temporally and spatially."[14]

To elaborate further on the phenomenological ground of the human-technology relation and its various trajectories, Don Ihde develops the generative concept of the relation from Martin Heidegger's earlier claim that the relation *per se* is ontologically primary.[15] That is, for Heidegger the term relational ontology means that objects and subjects are not separate entities which are then *a posteriori* gathered up in a relation; instead, the relation itself is *a priori*, such that our separation of subject and object, human and technology, nature and culture—or indeed this applies across the whole range of oppositional couplings—is heuristic and after the fact. With this understanding of relationality in mind, Ihde differentiates between various human-machine-world relations, each of which indicates a particular mode of *being* in relation to equipment. These relations are based on the correlational schema Human <==> World: the upper line of the bi-directional arrow indicates the "first intentionality" or the directedness of experience towards the world, while the lower "'reflective' intentionality is the movement from that which is experienced towards the *position* from which the experience is had. . . . [precluding] any simple talk about 'objects themselves' or, equally, 'subjects themselves,' since whatever falls out in the analysis is about *relations* between experiencers and experienced."[16]

Ihde categorizes these relations as follows: *embodiment* relations (semi-transparent relations such as wearing eyeglasses, where we experience the world through technology); *hermeneutic* relations (semi-opaque relations which require us to "read" a device such as a thermometer or gauge); and *alterity* relations (technology as "other" such as the computer or artificial intelligence). Also, in a more marginal and perceptually partial sense there are *background* or *field* relations (experience *among* technologies such as the constant background hum of domestic appliances), and *horizonal* relations (technology as surrounding and world-changing potential, such as the threat of nuclear holocaust or ozone depletion, an awareness which remains on the horizon of our collective consciousness). Each of these relations intersects in various ways with mobile phone use, but it is the embodiment relation and the background or field relation that is of most significance for location-based and casual gaming. The first describes how we "see the world" through the mobile screen as it focuses and dilates our sensory schemata, while the second indicates the way in which "we have our being among machines."[17]

In his phenomenological study of the spatial experience of telephone use, Robert Rosenberg usefully deploys Ihde's notion of the embodiment relation in complex combination with a type of background relation he terms *field composition*. This term describes, "a technology's potential to reorganize the overall structure of one's field of awareness as the technology is used."[18] Borrowing from Aron Gurwitsch, he suggests that as a technology becomes embodied, we experience a change to our *theme* (that which at any given moment is of central concern or a focus of attention), our *thematic field* (the experiential context of the theme), and that which resides at the margins or withdraws from awareness. This reorganization, Rosenberg argues, "involves a specific relation to the immediate space surrounding the user's body" but also reorganizes our perception and experience of the broader situation and the boundary between awareness and non-awareness.[19] Such experiences become conceptually and perceptually "sedimented" with habitual or routine use; as I discuss, such a sedimentation is occurring with the emergence of what Adriana de Souza e Silva and Eric Gordon call "net-local" public space,[20] increasingly afforded by location-aware mobile devices.

THE CORPOREALIZATION OF SPACE

Dean Chan makes the insightful observation that the "tension between the virtual and the actual resides at the heart of attendant debates about the modalities of co-presence in mobile telephony."[21] Such debates consider, at a fundamental level, the ontology of online and offline spaces and places and how we can be said to *be* in such environments. Many discussions of locative mobile gaming have problematized this distinction, suggesting that we simultaneously cohabit online space and physical place in our engagement with the mobile web as it intersects with peripatetic or pedestrian movement. Such debates challenge the notion that *dis*embodiment is a condition of using the internet or the phone, consider the ways that tele-technologies modify the body, and the kind of embodiment afforded by telepresent and mobile media. As de Souza e Silva and Daniel Sutko note, early discussions about cyberspace and the internet clearly demarcated the virtual from the "real" world, resulting in a "conceptualization of the Internet and digital information as essentially disconnected from physical spaces."[22] Thus, for example, in her well-known book *The Pearly Gates of Cyberspace*, Margaret Wertheim made the claim that "(t)he ontology of cyberspace is *ex nihilo*."[23] In her objectivist account, the virtual world is a colonizable dataspace that is vacant prior to our own informatically enframed "arrival" as cursor or avatar. According to Wertheim, the nature of virtual being, of what can be said to exist in cyberspace, has emerged as it were *out of nothing*, because space is an empty container prior to our occupation of it.

In everyday life, we tend to treat space as a constant, as an empty place or as a container for physical things. In this naturalized version of space, tools such as maps and the Cartesian coordinate system are just a way to measure the distance, size and location of things in space. Yet from a phenomenological perspective, our understanding of the geopolitical organization of space, of the globe and outer space, or simply of the extension between particular points, are all worked over by means of particular devices and technologies, and thus, in Ihde's sense, corporeally embedded in human-technology relations. The spatial representation of virtual spaces discloses our reliance on familiar and habitual corporeal models of "holding places" and our traversal within them. In a generic sense, the way in which we understand and navigate spatiality, and use spatial models of touring, mapping, topology and geometry, to "locate" ourselves in virtual space is necessarily and essentially in collaboration with modes of embodiment. Quite clearly, "spatial or 'orientational' metaphors are the most common of all. . . which has to do with the fact that mental mapping is 'grounded' in fundamental bodily experiences (our perceptions of back, front, beside etc). Spatial metaphors arise 'from the fact that we have bodies of the sort we have and that they function as they do in our physical environment'."[24] As Mark Hansen comments, for Merleau-Ponty the body and space are dynamically coupled, such that changes in bodily motility (e.g., walking while texting or using a navigational app) necessarily correlate with changes in lived spatiality, the sum of which is expressed in the body schema.[25]

Mobile devices clearly work to challenge any notion of a disembodied telepresence that is seemingly endemic to digital screen media, as we are frequently on-the-move, on-the-street and purposefully situated in local spaces and places when engaged in mobile phone use and mobile gameplay. For de Souza e Silva, location-aware mobile media and the increasing use of navigational and locative applications effectively interleave the physical and the digital, dissembling the dualism as both come together in "the immanence of the real."[26] Yet it is not that location-based mobile media applications create a condition where the disembodied virtual becomes re-embodied in located, situated, peripatetic practices, but rather that a different mode of embodiment is realized that conjoins the *corporeally realized* space of online networked interaction with pedestrian mobility. As Graeme Kirkpatrick suggests, each digital interface is a new "matter-form" that depends on the innovation of new material devices (such as portable game consoles or controllers), each defining "new spaces for the playful body to inhabit."[27] So in the case of mobile gaming, it is this embodied experience of the game space (whether location-based or casual) that becomes corporeally familiar via the remediation of previous interfaces and at the same time complexly embedded in pedestrian movement and the contextures of the built environment.

Neither is it the case that the "space" of online or distal interaction is a corporeal dilution of the "thick co-presence" of face-to-face interaction.[28] Theorists such as John Urry, Edward Casey and Paul Adams have argued that primarily visual and mediated experiences of any kind are sensorially impoverished or "thinner" than unmediated experiences such as walking which afford us with "a peripatetic sense of space."[29] Adams states: "The cultural transformations attending one technological diffusion, the diffusion of vehicles, parallel the cultural transformations linked to other technological diffusions. Computer simulations, video games, World Wide Web experiences, and other "virtual worlds" are overwhelmingly visual, and in their foundation in instantaneous connection they both reflect and reinforce the dislocated worldview of the driver."[30] Against this notion of a distinct hierarchy of sensorial quality between "unmediated" and "mediated" experiences, from a phenomenological perspective both are equally "multisensory," and vision must always already include the other senses. As Rowan Wilken and Richardson have argued, the very notion of haptic or tactile vision which can be richly described as an aspect of mobile phoning and gaming, discounts the view that perception isn't always already synesthetic.[31] Indeed, the techniques of haptic vision—where eyes and touchscreen coalesce—that are embodied by the mobile phone pedestrian point to the impossibility of distinguishing between a "peripatetic sense of space" on the one hand, and a mediated experience of the urban environment, on the other.

HYBRID SPACE AND "A SPACE-OF-ONE'S-OWN": LOCATION-BASED AND CASUAL MOBILE GAMING

Mobile media interfaces—or more precisely, for the purposes of this chapter and anthology, touchscreen web-capable devices that enable, among other things, both discrete and networked gaming—can be critically understood as complex and divergent instantiations of new media forms, each demanding a particular mode of embodied interaction. That is, the particular way we engage with mobile screens in public places determine (and are determined by) degrees of attention, practices of viewing and the motility and mobility of the body in situational contexts. The prioritization of modes of mobile phone use (casual or network gaming, listening to music, watching TV, film-making and editing, photography, web browsing, gaming, video-phoning, texting and media-messaging) reflect different relationships between users, bodies, content, handsets and the physical environment or spatial context.

As such, previously discrete media functionalities come together and are mobilized in the one device—in the case of the iPhone, internet browsing, downloadable apps, locative functionality, haptic touchscreen, casual and networked modes of interaction and gaming—and what emerges is not a unitary interface but a range of activities that each prioritize a specific *technosomatic* arrangement. If, as Eikenes and Morrison would suggest, the mobile interface is a multimodal site of activity, it is salient to examine the perceptual specificity of such activities, and for the purposes of this chapter, to acknowledge the ways in which particular mobile games (casual, location-based, hybrid, online, discrete, etc) reflect different relationships between users, handsets, content, the physical and socio-cultural environment, and spatial and corporeal circumstances. Different kinds of content and game modalities literally change "what there is to be seen,"[32] how things appear and what we attend to, effecting the way in which we interact with others, and how we move within and experience urban space and place. Mobile gameplay expresses not only a "way of being" in the world, but also a "way of being together" that requires mutual spatial and corporeal adjustment.[33]

It is in this context that I offer a comparative analysis of location-based gaming and casual gaming with specific reference to the iPhone. On a macro-perceptual level, in the case of location-based gaming which integrates the player's perception and experience of the built environment with GPS data and real-time navigational apps, our bodies are mapped both in physical and online space in a complex and interspersed layering. In early 2010, for example, 200 *Crimsonfox* players, having downloaded the *Shibuya Scanner* iPhone app that provided a camera overlay showing GPS information using the built in iPhone compass, ran around the city searching for hints that would reveal the "real-world" hideout of the Moonlight secret society.[34] At the micro-perceptual level, in its deployment of a multi-touch interface, accelerometer, GPS, real-time 3D graphics and 3D positional audio, the iPhone demands a unique *corporealization* of gameplay. Moreover, in enabling game and application developers to create and upload their own creations into the App Store, there is potential for significant innovation and variation in terms of the technosomatic skills required by the gamer. The game *QWOP*, for example, requires the player to use two fingers or thumbs to awkwardly manipulate a sprinter's individual leg muscles like a puppeteer across the iPhone screen; designed by Bennett Foddy as a way to explore how we embody screen characters, the game is frustratingly difficult and requires "the most intense micromanagement" of the hand and dedicated focus on the screen, in order to overcome the "deliberate disconnect" between intention and action.[35] These brief examples give some indication of the variable corporeal schemata required for different kinds of mobile gameplay; in what follows, I turn to a more detailed analysis of location-based and casual gaming.

Location-based Mobile Games

As de Souza e Silva and Jordan Frith note, both the GPS-enabled iPhone and Google's Android operating system "contributed to the popularization and commercialization of location-aware applications" and location-based services, which typically provide situational information about the urban environment via online databases and media libraries, such that informational changes on the mobile screen effect the navigation and experience of physical space.[36] A number of theorists have argued that the proliferation of mobile online activities—via mobile phones, laptops, PDAs and more recently netbooks and iPads—are changing the way we think about being "on", "at" or "in" a simulated or computer space, and the way we think about being "on" or "off" line.[37] As de Souza e Silva claims:

> Because many mobile devices are constantly connected to the Internet
> . . . users do not perceive physical and digital spaces as separate enti-
> ties and do not have the feeling of "entering" the Internet, or being
> immersed in digital spaces, as was generally the case when one needed
> to sit down in front of a computer screen and dial a connection.[38]

Being online and networked thus becomes another function of the mobile
phone, but it is importantly a *different* experience of the internet and online
connectivity: being online becomes enfolded inside present contexts and
activities, like the embodied and itinerant acts of walking, driving, face-
to-face communication and numerous other material and somatic involve-
ments. Location-based mobile gaming is theorized as a particularly robust
example of this emergent hybrid ontology.[39] In his analysis of *Geocaching*, Jason Farman describes the mixed or augmented realities of pervasive
location-based games where bodies, networks and material space con-
verge.[40] Played in over 200 countries, *Geocaching* is treasure hunt game
requiring game players to hide "geocache containers" marked with GPS
data in public places; players then "use their mobile devices (from GPS
receivers to iPhones) to track down the container, sign the log, and leave
tradable and trackable items in the cache."[41] In such games, our embodied
proprioception—that is, the awareness of our body's position in relation
to the environment enabled by our perception of movement and spatial
orientation—must seamlessly accommodate both immediate and mediate
being-in-the-world. Players of location-based games and users of location-
based services, Farman comments, navigate the landscape in a "simultaneous process of sensorial movement through streets and buildings and an
embodied connection to how those places are augmented by digital information on mobile devices."[42] Similarly, Licoppe and Inada describe players
of the game *Mogi*[43] as "hybrid beings" who are able to "smoothly integrate
the embodied lived experience of the body and the mediated perception of
oneself and of the environment."[44] For Farman, this mergence increasingly
constitutes the "interface of everyday life."[45]

There are a number of iPhone apps that exemplify this interleaving
of online and physical environments. *ARIS*, a web-based authoring tool
with an "ultra-permissive license" available as an iPhone App, allows
users to make mobile games, tours and interactive stories. One example
of an *ARIS*-based user-generated experience is *Bike Box*; participants
pedal around central Brooklyn, and upload site-specific audio through the
iPhone app, as they listen to a "curated collection of geo-specific sounds
provided by a variety of local land-use experts, historians, poets, artists, and other interpreters."[46] *Bike Box* creators aim to give users "access
to the layers of lived experience, personal anecdote and history that are
piled up invisibly on every street corner and city block."[47] More recently,
MMO game *Kiss My Rocket* locates gamers geospatially and uses that
data to create a network space of global conflict; players locate "enemies"

anywhere in the world via an interactive world map (with 12 zoom levels!) and fire missiles in real time at the bases of other players.[48] *MyTown*, a location-based social game reportedly played by over 3.3 million people, enables players to "virtually" buy their favorite stores and places (at the same time earning "cash" to spend in real locations) turning the "real world" into a Monopoly game.[49]

Such games and applications and their deployment within urban space means that we need to rethink the spatial and place-based experience of being-in-public. That is, the use of location-based mobile apps literally generate hybrid spaces by integrating online information of one's immediate environment into the patterns of urban life and peripatetic movement. In recent work de Souza e Silva and Gordon have argued that such hybrid practices generate what they term *net-local public space*, which describes our movement "between the immediately proximate and the mediately distant within a carefully crafted set of social rituals that ultimately serves to extend the purview of local space."[50] Net-local public space includes those engaging in location-based activities with mobile devices, those (both co-present and online) inhering or participating in this network activity, and those non-participants who are co-located in the urban setting. As our "attentional foci" in such quotidian spaces becomes diversified and hybridized, effecting new micro- and macro-movements, the actual/virtual dichotomy previously used to differentiate between offline/online practices is thoroughly disassembled into a complex and dynamic range of modalities of presence. Yet it might be said that even prior to the emergence of location-based and hybrid reality gaming, mobile and portable devices instantiated hybrid experiences and ontologies in various forms. As N. Katherine Hayles[51] notes, the mobile phone works to "enfold" contexts, such that urban spaces are now filled with mobile phone users who create communicative "pockets" of co-existing modalities of co-presence, telepresence, absent presence, distributed presence and ambient presence, all of which demand different modes of embodied being-in-the-world.

As Giusepppina Pellegrino comments, hybridity "is the key-word which, more than others, describes the co-constructed dimension of participation" in contemporary media culture.[52] It is worth noting here that there are several ways we can "think" hybridity. De Souza e Silva and Pellegrino use the term to refer to the way embodiment is transformed as an effect of "changes in forms of co-presence," when "participation goes beyond physical co-presence and is experienced through multiple forms of proximity, both physical and virtual"[53]; here, presence does not mean being "being-there" but rather what we experience is a "soft relationality" or the layering of several modalities of presence.[54] In this way, location-based mobile applications can be said to add a complex *dimensionality* to place and space. Yet in a phenomenological sense, hybridity more simply describes the relational ontology of bodies, technologies and the affordances of the environment. The human condition is always-already one of relationality and hybridity;

thus, in a fundamental sense, the experience of hybridity is one to which we are always already corporeally well-attuned.

Indeed, this ability to embrace "fresh instruments" and mediated perception and modes of embodiment within one's corporeal schema, and to oscillate between, conflate and adapt to ostensibly disparate modes of being and perceiving, is precisely why telepresence and virtual space are both somatically and ontologically tolerable. We might say that an "as-if" structure of presence and mobility is fundamental to our experience of the hybrid space of location-based gaming—the terms mobility and presence must account for the physical macro-movement of the pedestrian body which can be traced geospatially, the micro-movements and motor coordination required of the mobile player as they negotiate screen-space, and the virtual movement and exchange of objects and creatures "into" the gamers' mobile devices and their passage through the hybrid game-space. Location-based mobile phone games thus potentially work to seamlessly combine the corporeal schematics of actual and virtual worlds as they are actively negotiated on-the-move, effectively creating a hybrid mode of being where the boundary between game and real life collapses.

Casual Mobile Games

Casual mobile gaming is often characterized as a mode of engagement that requires only sporadic attention up to a threshold of around five minutes, hence the popular notion that casual games are the mobile phone's predominant game genre, and the labeling of casual gamers, who play at most for five minutes at a time and at irregular intervals, as a key market in the mobile game industry. Casual gamers are deemed to deliberately avoid the corporeal attachment of dedicated console or PC gameplay so that they are perpetually ready to resume their temporarily interrupted activities. Yet in *A Casual Revolution* Jesper Juul[55] suggests that the stereotypes of the hardcore and casual gamer over-simplify the often complex and variable modalities of play, and that the recent proliferation and popularity of casual games (such as classic puzzle games) reach across many demographics. Such games are typified by "interruptability," where play becomes intertwined with everyday routines, and "fit" into the existing patterns of life.[56] Similarly, Kirsi Kallio, Frans Mäyrä, and Kirsikka Kaipainen stress the importance of the "variability of meanings" attached to gameplay, the "situatedness and contextuality of gaming" and "the layered and overlapping character of game mentalities," which are particularly relevant now that gaming is permeating everyday spaces and cultures.[57] For example, Chan cites a public survey taken in 2006 which found that, contrary to the common perception that casual mobile games are played while on-the-move, for most Japanese casual mobile games are played in their bedroom or in the home, suggesting that mobile games are often engaged with in non-mobile, sometimes private and "sedentary domestic contexts."[58] Yet even home-based casual mobile gaming is too broad a descriptor for the kinds of gameplay that take place; as Kallio, Mäyrä, and Kaipainen have argued, there are more fine-grained differences between "killing time" (varying between concentrated long-term and less focused short-term play mixed in with other activities), "filling gaps" (taking a break between activities, filling in an empty moment, moving from one task to another, such that the length of gameplay is dependent on the next task or activity) and "relaxing" (when there's no "work" to be done and gameplay can extend for hours).[59]

In phenomenological terms, casual mobile gaming on touchscreen and motion-sensing devices can at times demand a non-casual multi-sensory engagement, perhaps more akin to the stickiness of console gaming in Chris Chesher's terms, or at least comparable to DS and PSP gameplay.[60] That is, traditional console and computer games (adventure, racing, first-person shooter etc) played on the iPhone (e.g., *Mario Racing* or *Nitro-Cart*) demand a corporeal attachment that necessitates an adroit oscillation between game-space and attentiveness to one's spatial surroundings. In this way, the iPhone screen could be said to challenge the perception of mobile games as predominantly casual or *nagara* games (i.e., played while doing something else), recuperating some of the adhesive and immersive qualities proper to console games played on a television or computer screen. Or, at least, we might identify a broad spectrum of attachment across a range of mobile gameplay—from casual games to location-based games—based on levels of immersion, engagement and distraction.

In part, the private space-making, cocooning and stickiness of some modes of casual mobile gaming have to do with the level of sensory immersion afforded by haptic touchscreens such as the iPhone—devices that have what is called a post-WIMP interface.[61] The iPhone is specifically created for use with the finger or fingers for multi-touch sensing, and because the screen is a capacitive touchscreen, it depends on electrical conductivity that can only be provided by bare skin. Moreover, the iPhone screen can track the movement of five fingers simultaneously; in a keynote speech on game development for the iPhone, *Freeverse* game designer Justin Ficarotta stated: "Several touches can be combined into gestures . . . Drags, swipes, flicks, pinches, with a variable number of fingers . . . It's very different from what we're used to with mainstream games."[62] Post-WIMP interfaces thus require a somatic and visceral understanding of naïve physics; for example, primary bodily "sensations" such as inertia and springiness can be found in many iPhone's applications and games and provide the synesthetic illusion that windows, objects and icons on the device have mass.[63] Naïve physics can also include our body-memory of hardware such as the keyboard and joystick that are simulated in the iPhone GUI.

Thus, there is a certain haptic intimacy that renders the iPhone an object of tactile and kinesthetic familiarity, a sensory knowing-ness of the fingers that correlates with what appears on the small screen. This is nowhere better exemplified than in iPhone games that depend on reality-based interaction which enfold the player into a temporary and incomplete simulation of real-world physics. The reality-based features of the iPhone (the multi-touch interface, the accelerometer, GPS, real-time 3D graphics and 3D positional audio) are deployed in a number of racing games where the iPhone device simulates a steering wheel, and also in games such as *MonkeyBall* and *Labyrinth,* which requires the player to tilt the screen on a horizontal plane to control the movement of the ball around holes and through various obstacles and gradients.

For Eikenes and Morrison,[64] as we come to habitually use such interfaces, we effectively develop a new kind of "motion literacy" specific to devices that combine touchscreen with accelerometer or position-recognition functionality such as the iPhone; a type of kinetic and motile learning is required—or in phenomenological terms, the appropriation of a "fresh instrument" into our corporeal schema—that works to overcome or adapt to the imprecise control we have over objects and actions in and on the screen. This is possible because of our ability to take on an "as if" structure of embodiment. Eikenes and Morrison coin the term *navimation* to describe "the intertwining of visual movement with activities of navigation in screen interfaces" and the way we can rotate screen orientation and push objects across the iPhone screen *as if* we "share gravity" with the device.[65] They comment that "there is no one-to-one mapping between physical and virtual space" (i.e., there is both a delay between our actions and the virtual rotation of the screen, and the iPhone screen "environment" can only be locked in four ways), yet regardless of this disjunction between real space and screen space, we easily fill in the gaps to achieve a satisfying kinetic response.[66]

As Jeff Rush notes, the embodied metaphor as it is deployed in gameplay—a trope that engenders "a heightened sense of the linkage between two different orders of reality, real physical gesture and its on-screen representation"—works to attach a "kinetic materiality" to the action and movement that takes place on the screen, creating moments of tangibility and concreteness.[67] Thus, for example, in the game *Paper Toss* (where the player "flicks" a piece of crumpled paper into a bin in a simulated office space), there is something of the kinetic experience of tossing paper that effectively becomes "condensed into the hand."[68] In part, this is achieved by what Paul Skalski et al. call *kinesic natural mapping,*[69] where bodily movement corresponds in an approximate (or "as-if") way to on-screen action, an effect enabled by the way touchscreens can deploy physical analogies; natural mapping works to "complete" being in a mediated space, facilitating an immersive experience. The kinetic experience is also achieved or augmented by synesthetic effect, and exploits our phenomenological ability to "perceive and integrate information from different modalities into the one complete sensation"[70]; thus, the sounds that accompany haptic games such a *Paper Toss* and *Fly Fishing 3D* (where the user imitates the action of casting a fishing line by "casting" the iPhone) effectively simulates (or stimulates) tactile feedback and increases the sense of being-in a discrete, tangible and sticky game-world.

As Hjorth and Richardson[71] have suggested, the activity of casual gaming in urban space also frequently takes place while waiting (for a friend, at a bus stop, or for a journey to end) and becomes a way of managing the corporeal agitation of impatience, aloneness and boredom in public spaces, while at the same time maintaining an "environmental knowing," or crucial peripheral awareness of one's spatial surroundings in readiness for the busy-ness of life to resume. The mobile device becomes co-opted into the corporeal labor of waiting, filling and suturing the "dead" or "fractured" times and spaces that are "folded into everyday corporeal existence."[72] Such work—categorized by what David Bissell refers to as the various "species" of waiting—can be understood in terms of the "micro-bodily actions" and "corporeal attentiveness" of specific modalities of waiting.[73] Yet such modalities of gameplay, though intermittent, enact a quite different experience of space and place in urban environments than that of location-based gaming, an interiorized distractedness and exclusionary mode of being-alone-together, where the game is experienced as a "little world" contained in the device. That is, in the case of casual mobile gaming in urban contexts, the perceptual fields of in-game and outside-world can often be still quite distinct, reminiscent of previous modalities of being-alone-together on public transport or other "waiting places" characterized by modernity.

For Jussi Parikka and Jaakko Suominen,[74] the "third place" between public and private space opened up by the mobile phone—in particular, via the use of mobile entertainment services, games, music and videos—demarcates a privatized space around the user, an habitual practice already common in the nineteenth century. They write:

> [W]hat is new in this division of space and creation of a place of one's own? Instead of seeing this solely as a trend of digital mobile culture, we argue that this is more a phenomenon that took off with the creation of modern urban space and the new paradigms of media consumption. . . [T]he pattern of mobile entertainment usage as the creation of a private sphere was already part of the railway culture of the nineteenth century—even if people consumed such media content as newspapers and books instead of *digital* entertainment.[75]

Thus, rather than a modality of gameplay that opens a hybrid space coalescing urban environment and online networks, some forms of "discrete," offline, casual mobile gaming can be seen as a form of portable home entertainment that assists us to achieve occasional seclusion when in public. It is this closing-off that prompts Stephen Groening to comment that a society of "portable personal electronics is a society in which private space is as physically mobile as the populace and privacy itself is radically mobile."[76] As Hjorth argues (see her chapter on the "iPersonal" in this volume) the mobile phone can frequently be experienced as a micro-mobile home, effecting a mobilization of private space.

Similarly, Michael Bull suggests that portable sound-based technologies such as the Walkman, mobile phone, iPod and MP3 player have contributed (along with the automobile) to the transformation of the urban soundscape by way of an auditory privatization of public space.[77] I have argued elsewhere[78] that while the iPod or MP3 player provides a continuous sound-bubble or "sonorous envelope" that effectively allows the user to deny the contingencies of the outside world,[79] the mobile phone is experientially discontinuous, "puncturing" the soundscape via the sporadic and unpredictable contingency of unexpected calls and text messages. The mobile music player is thus discrete and cocooned, whereas the mobile phone user "colonizes" urban space, intermittently carving out a place of communication and telepresent intimacy, temporarily irrupting the immediate soundscape with personal ringtones, bleeps and one-sided conversations. Yet the mobile offers diverse way of being-with-others; mobile casual gaming, when deployed as a way to create a "space of one's own" in public spaces, returns us—albeit sporadically—to the practice of mobile privatization. We can adeptly choose—and spontaneously oscillate between—different levels of attention, inattention and distraction when casual gaming.

CONCLUSION

In this chapter, I have sought to describe location-based and casual mobile gaming in terms of the way they are *intercorporealized* within the patterns and contextures of everyday life. Using a phenomenology-informed approach, I have deliberately focused on these two quite specific modalities of mobile gameplay as somatically and spatially distinct experiences, in order to illustrate how the complexities of mobile media practices *en large* are continuing to unfold. On the one hand, location-based gaming effectively conjoins our pedestrian movement and placement in urban environments with geospatial data and online social networks, such that we engage in a kind of "hybrid-space walking" that transforms the way we experience and "attend to" both the city and being together.[80] On the other, casual mobile gameplay can be used to demarcate a discrete "space-of-one's-own" in busy public contexts, and often demands an intense micro-perceptual closed-circuit between eyes, hands and screen. Each mode of gaming activates "different kinds of vectors of movement and rest, sociability and individuality"[81] and transforms the way we experience our "being" and "doing" in the world. Throughout my analysis I have also, at various points, considered the iPhone and its particular affordances—capacitive touchscreen, application marketplace, and a web browser adapted to the "already-mature ecosystems of the wired web"[82]—as both a lens through which to examine an emergent human-technology relation, and as an exemplar of mobile media and the ludic turn in contemporary technoculture.

NOTES

1. William Saletan, "The mind-BlackBerry problem: Hey, you! Cell-phone zombie! Get off the road!" *Slate,* http://www.slate.com/id/2202978. Cited in Jason Kalin, "Toward a Rhetoric of Hybrid-Space Walking", in *Proceedings of the Media Ecology Association*, v.10, June 19–21 2009, Saint Louis University http://www.media-ecology.org/publications/MEA_proceedings/v10/6_Hybrid_space_walking.pdf, 50 (accessed 8 May 2011)
2. Jon Olav H. Eikenes and Andrew Morrison, "Navimation: Exploring Time, Space & Motion in the Design of Screen-based Interfaces", *International Journal of Design*, 4(1), 2010: 1–16 (1).
3. Joel West and Michael Mace, "Value creation in the mobile internet: the impact of Apple's iPhone," 14 February 2008, http://www.joelwest.org/Papers/WestMace2008.pdf (accessed 10 April 2011).
4. Joel West and Michael Mace, "Browsing as the killer app: explaining the rapid success of Apple's iPhone," *Telecommunications Policy* 34(5–6), 2010: 270–286.
5. Ibid., 280.
6. Gerard Goggin, "Adapting the mobile phone: the iPhone and its consumption," *Continuum* 23(2), 2009: 231–244.
7. West and Mace, "Browsing as the killer app," 282.
8. Ibid., 270.
9. Ibid., 279.
10. West and Mace, "Value creation."
11. Robert F. Nideffer, "Game engines as open networks," in Joe Karaganis, ed., *Structures of Participation in Digital Culture* (New York: Social Science Research Council, 2007), 200–217.
12. Maurice Merleau-Ponty, *Signs* (Evanston, IL: Northwestern University Press, 1964).
13. Ibid. , 5.
14. George Simmel, "Metropolis and mental life," in Gary Bridge and Sophie Watson, eds., *The Blackwell City Reader*, (Malden, MA: Blackwell Publishing, 2002), 11–19. Cited in Jason Kalin, "Toward a rhetoric," 54.
15. Don Ihde, *Technics and Praxis* (Dordrecht, Holland: D. Reidel Publishing Company, 1979).
16. Ibid., 17.
17. Ibid., 15.
18. Robert Rosenberg, "The spatial experience of telephone use," *Environment, Space, Place* 2(2), 2010: 61–75.
19. Ibid., 69.
20. Adriana de Souza e Silva and Eric Gordon, "Net-local public spaces: towards a culture of location," unpublished paper, http://www.urbancomm.org/dynamic_images/seminars/ (accessed 5 August 2010).
21. Dean Chan, "Convergence, connectivity, and the case of Japanese mobile gaming," *Games and Culture* 3(1), 2008: 13–25.
22. Adriana de Souza e Silva and Daniel M. Sutko, "Theorizing locative technologies through philosophies of the virtual," *Communication Theory* 21(1), 2011: 25.
23. Margaret Wertheim, *The Pearly Gates of Cyberspace: A History of Space from Dante to the Internet* (London: Virago Press, 1999), 221.

24. Jostein Gripsrud, *Television and Common Knowledge* (New York: Routledge, 1999), 119.
25. Mark Hansen, "Embodying virtual reality: touch and self-movement in the work of Char Davies," *Critical Matrix: The Princeton Journal of Women, Gender and Culture* 12(1–2), 2001: 112–147.
26. de Souza e Silva and Sutko, "Theorizing locative technologies," 34.
27. Graeme Kirkpatrick, "Controller, hand, screen: aesthetic form in the computer game," *Games and Culture* 2009: 127–143.
28. John Urry, "Mobility and proximity," *Sociology* 36(2), 2002: 255–274.
29. Ibid.; Edward S. Casey, "Between geography and philosophy: what does it mean to be in the place-world," *Annals of the Association of American Geographers* 91(4), 2001; Paul C. Adams, "Peripatetic imagery and peripatetic sense of place," in Paul C. Adams, Steven D. Hoelscher and Karen E. Till, eds., *Textures of Place: Exploring Humanist Geographies* (Minneapolis: University of Minnesota Press, 2001), 186–206.
30. Adams, "Peripatetic imagery," 191.
31. Ingrid Richardson and Rowan Wilken, "Haptic vision, footwork, place-making: a peripatetic phenomenology of the mobile phone pedestrian," *Second Nature: International Journal of Creative Media* 2(1), 2009. http://secondnature.rmit.edu.au/index.php/2ndnature/article/view/121/35
32. Jean-Paul Thibaud, "Sensory design: the sensory fabric of urban ambiences," *Senses and Society* 6(2), 2011: 203–215.
33. Ibid., 210.
34. Einat Cohen, "Portable gaming in Japan: redefining urban play space and changing gameplay," University of Haifa, May 2010, http://east-asia.haifa.ac.il/semminar-best-works/gampeplay_final.pdf (accessed 7 April 2011).
35. *Mark Brown*, http://www.wired.com/gamelife/2011/03/qwop-girp/ (accessed 29 March 2011).
36. Adriana de Souza e Silva and Jordan Frith, "Locative mobile social networks: mapping communication and location in urban spaces," *Mobilities* 5(4), 2010: 485–506
37. Frank Lantz, "Big games and the porous border between the real and the mediated," *receiver magazine* 16, 2006, http://www.receiver.vodafone.com/ (accessed 18 November 2007).
38. Adriana de Souza e Silva, "From cyber to hybrid: mobile technologies as interfaces of hybrid spaces," *Space and Culture* 9(3), 2006: 261–273.
39. Ibid.; Jason Farman, "Locative life: geocaching, mobile gaming, and embodiment," *Proceedings of the Digital Arts and Culture Conference— After Media: Embodiment and Context,* University of California, Irvine, 12–15 December 2009, http://escholarship.org/uc/item/507938rr (accessed 10 March 2010); J. Follett, "The world as the interface—location data and the mobile web," *Receiver Magazine* 21, 2008, http://www.vodafone.com/receiver/21/articles.html (accessed 8 December 2009); Marinka Copier, "Challenging the magic circle: how online role-playing games are negotiated by everyday life," in Marianne van den Boomen, Sybille Lammes, Ann-Sophie Lehmann, Joost Raessens, and Mirko Tobias Schafer, eds., *Digital Material: Tracing New Media in Everyday Life and Technology* (Amsterdam: Amsterdam University Press, 2009), 159–172.
40. Farman, "Locative life."
41. Ibid.
42. Ibid.
43. In *Mogi,* the city is represented both as a map on players' mobile phones and on the web, the latter of which provided online players with an expanded view of the gamespace overlaying Tokyo along with the geographic and

gameworld location of all players. Mobile and online players both have different "views" of the gamespace, and collaborate to collect virtual objects and creatures at various locations throughout the city. It is this collaboration that "constructs" the hybrid space. See Adriana de Souza e Silva and Larissa Hjorth, "Playful urban spaces: a historical approach to mobile games," *Simulation & Gaming* 40(5), 2009: 602–625.

44. Christian Licoppe and Yoriko Inada, "Emergent uses of a multiplayer location-aware mobile game: the interactional consequences of mediated encounters," *Mobilities* 1(1), 2006: 39–61.
45. Farman, "Locative life."
46. *ARIS*, http://arisgames.org/featured/the-bike-box/ (accessed 7 June 2011).
47. Ibid.
48. *Kiss My Rocket*, iTunes App Store, 15 July 2011.
49. Ben Harvel, "Location-based iPhone games growing in popularity. *MyTown* creator completes $20m round of financing," *148Apps*, 18 May 2010, http://www.148apps.com/news/locationbased-iphone-games-growing-popularity-mytown-creator-completes-20m-financing/#ixzz1SbuqY42M (accessed 7 June 2011).
50. de Souza e Silva and Gordon, "Net-local public spaces."
51. N. Katherine Hayles, 2002, personal communication, cited in de Souza e Silva, "From cyber to hybrid," 28–29.
52. Giusepppina Pellegrino, "Mediated bodies in saturated environments: participation as co-construction," in Leopoldina Fortunati, Jane Vincent, Julian Gebhardt, Andraz Petrovcic and Olga Vershinskaya, eds., *Interacting with Broadband Society* (Frankfurt am Main: Peter Lang, 2010): 93–105.
53. Ibid., 99.
54. Chris Speed, "Developing a sense of place with locative media: an 'underview effect'," *Leonardo* 43(2), 2010: 169–174.
55. Jesper Juul, *A Casual Revolution: Reinventing Video Games and Their Players* (Cambridge, MA: MIT Press, 2010).
56. Ibid.
57. Kirsi Pauliina Kallio, Frans Mäyrä and Kirsikka Kaipainen, "At least nine ways to play: approaching gamer mentalities," *Games and Culture* 6(4), 2011: 327–353.
58. Chan, "Convergence," 23.
59. Kallio et al., "At least nine ways to play," 342.
60. Chris Chesher, "Neither gaze nor glance, but glaze: relating to console game screens," *SCAN: Journal of Media Arts Culture,* 1(1), 2004, http://scan.net.au/journal/ (accessed 20 August 2009).
61. WIMP stands for Window, Icon, Menu, Pointing Device, the standard tools for navigation and control of the interface. Robert J. K. Jacob et al., "Reality-based interaction: a framework for post-WIMP interfaces," in *CHI 2008 Proceedings: Post-WIMP,* Florence, Italy, 5–10 April 2008: 201.
62. Jill Duffy, "iPhone game development tips," *Game Career Guide* (10 September 2008), http://www.gamecareerguide.com/features/624/iphone_game_development_tips.php (accessed 5 June 2010).
63. Jakob et al., "Reality-based interaction."
64. Eikenes and Morrison, "Navimation."
65. Ibid., 14.
66. Ibid., 12–13.
67. Jeff Rush, "Embodied metaphors: exposing informatic control through first-person shooters," *Games and Culture* 6(3), 2011: 245–258.
68. Kirkpatrick, "Controller, hand, screen," 134.

69. Paul Skalski, Ron Tamborini, Ashleigh Shelton, Michael Buncher and Pete Lindmark, "Mapping the road to fun: natural video game controllers, presence, and game enjoyment," *New Media & Society* 13(2), 2011: 224–242.
70. Even Hoggan, Topi Kaaresoja, Pauli Laitinen and Stephen Brewster, "Crossmodal congruence: the look, feel and sound of touchscreen widgets," *ICMI'08*, 20–22 October 2008, Chania, Crete, Greece.
71. Larissa Hjorth and Ingrid Richardson, "The waiting game: complicating notions of (tele)presence and gendered distraction in casual mobile gaming," *Australian Journal of Communication* 36(1), 2009: 23–35.
72. David Bissell, "Animating suspension: waiting for mobilities," *Mobilities* 2(2), 2007: 277–298.
73. Ibid., 278, 282, 285.
74. Jussi Parikka and Jaakko Suominen, "Victorian snakes? Towards a cultural history of mobile games and the experience of movement," *Game Studies* 6(1), 2006, http://gamestudies.org/0601/articles/parikka_suominen (accessed 8 January 2010).
75. Ibid.
76. Stephen Groening, "From 'a box in the theater of the world' to 'the world as your living room': cellular phones, television and mobile privatization," *New Media & Society* 12(8), 2010: 1331–1347.
77. Michael Bull, "Thinking about sound, proximity and distance in Western experience: the case of Odysseus's Walkman," in Veit Erlmann, ed., *Hearing Cultures: Essays on Sound, Listening and Modernity* (Oxford and New York: Berg, 2004).
78. Ingrid Richardson, "Audile telepresence: a sonic phenomenology of mobile phones," *ANZCA09: Communication, Creativity and Global Citizenship Conference*, Queensland University of Technology, Creative Industries Precinct, Brisbane, Australia, 8–10 July 2009.
79. Bull, "Thinking about sound," 185, 189.
80. Jason Kalin, "Toward a rhetoric of hybrid-space walking," 54.
81. Parikka and Suominen, "Victorian snakes."
82. Mace and West, "Browsing as the killer app," 271.

Part III
iPhone and Labor

10 The iPhone as Innovation Platform
Reimagining the Videogames Developer

John Banks

INTRODUCTION

In 2007, *Time Magazine* celebrated the launch of the iPhone by declaring it invention of the year. *Time's* Lev Grossman justified the accolade by stating, "It's not a phone, it's a platform."[1] This was his fourth of five reasons—the others were, "The iPhone is pretty"; "It's touchy-feely"; "It will make other phones better"; and "It is but the ghost of iPhones yet to come." In coming to this judgment, Grossman notes that Apple did not invent, or even reinvent, the touchscreen, "but Apple knew what to do with it." He observes, "Platforms are for building on. Last month, after a lot of throat-clearing, Apple decided to open up the iPhone, so that you—meaning people other than Apple employees—will be able to develop software for it too. Ever notice all that black blank space on the iPhone's desktop? It's about to fill up with lots of tiny, pretty, useful icons."[2] And today, these icons now include lots of videogame apps.

In this chapter, I consider the proposition that the iPhone provides an *innovation platform* for the videogames industry. What are the implications of framing the iPhone as an innovation platform? Does adopting an innovation framework reduce analysis to a narrowly instrumental rationality? Does it restrict us to considering technical and engineering indicators of R&D? If we widen our range of innovation indicators to include aesthetic novelty or organizational transformation does the term innovation start to lose analytic clarity and rigor? How is innovation organized? I explore these questions through a brief case study of HalfBrick, a Brisbane, Australia-based videogames developer that has enjoyed considerable commercial success with the release of game applications for the iPhone such as *Fruit Ninja,* which as of July 2011 was in second place on the Top Paid iPhone Apps list. This case study draws on the early stages of a research project, partnering with Australian interactive entertainment companies, that aims in part to explore and investigate via comparative ethnographic research the sources and processes of innovation in this sector of the creative industries.[3]

In drawing on this research, I focus on the professional identities of games developers as they re-engineer the very process of making games while innovating for mobile devices such as the iPhone. I suggest that the remaking of professional creative producer identity is core to questions about innovating for and through the iPhone. What kinds of "professional imaginary" are emerging to negotiate the opportunities and challenges of our digital platform working environments?

It is clear why the iPhone offers an attractive market opportunity for videogames developers. Even a cursory glance through the offerings of Apple's App Store suggests that the iPhone is a significant and growing platform for videogames. The games category regularly features in, and indeed dominates, the list of Top Ten Paid iPhone Apps. As of July 2011, the top four apps are games and there are eight game titles in the top ten. More broadly, the digital distribution of games, especially for mobile devices, is shaping up as a significant structural transformation of the videogames industry. Industry commentator Tim Merel suggests that the growing market for mobile and online social and casual games indicates a split in the industry with traditional console markets stagnating.[4] Nielsen recently reported that Apple's iPhone leads the mobile phone category for gaming and that 64% of the apps used in a 30-day period by app downloaders were in the games category.[5] This report, however, does not detail how this use was tracked. This research found that these patterns were especially pronounced on the iPhone, with iPhone owners using the device to play games far more often than owners of other devices (14.7 hours per month for iPhone users compared to 9.3 hours per month for Android users). The report also finds that app downloaders are more willing to pay for games than for any other type of application.[6]

The interactive affordances and haptic pleasures of the iPhone's touchscreen have given rise to a slew of arguably innovative titles including Rovio Mobile's phenomenally successful *Angry Birds*, the top-selling iPhone game. In *Angry Birds* players use a slingshot to fire birds at pigs situated on or within various structures with the aim to destroy all the pigs. By completing levels and advancing through the game, players "unlock" new birds with special abilities that the player can activate to solve various challenges. As of 15 July 2011, *Angry Birds* maintained its number one position in the App Store and now features some 225 levels, together with leader boards and achievements players can earn and display through Apple's Game Center support. *Angry Birds* has also been ported to other platforms including Android systems. The development team is currently working on a version of the game for Facebook.

Angry Birds' commercial success suggests that the iPhone provides an excellent platform opportunity for videogames developers. The iPhone offers developers access to a potentially new and growing market of casual game consumers who use their phones for playing games. However, aren't Apple's "walled garden" and App Store constraints and controls the very antithesis of the openness that we should encourage and value? And, in all of this, what is at stake in considering the status of the iPhone as a platform? Jonathan Zittrain's critique of the iPhone as a "tethered appliance"[7] characterized by networks of control rather than a generative technology is a strong example of such a perspective. He argues that "these appliances take the innovations already created by Internet users and package them neatly and compellingly, which is good—but only if the Internet and PC can remain sufficiently central in the digital ecosystem to compete with locked-down appliances and facilitate the next round of innovations."[8] However, in developing this critique Zittrain positions these "appliances" in opposition to more generative technologies, which he describes as "platforms."[9] For Zittrain, the iPhone was not generative precisely because it was not a *platform*. In responding to Zittrain, Gerard Goggin argues that through a process of adaptation, "the iPhone is now a more open platform, with a set of controls but also better access for developers, and also having fermented a thriving user culture."[10] To be fair to Zittrain, since the launch of the App Store in July 2008, the iPhone has become more of a generative platform with the growing availability of various third-party apps, including games. At the heart of this idea of platforms is the capacity for others to develop for the device. Marc Andreessen defines platform as:

> A system that can be reprogrammed and therefore customized by outside developers—users—and in that way, adapted to countless needs and niches that the platform's original developers could not have possibly contemplated, much less had time to accommodate.[11]

Platforms are central to the process of developing videogames: developers' project practices and the associated daily work of programmers, designers and artists are shaped by the affordances and constraints of the particular platform they are targeting—from PCs and consoles to mobile phones and social-network platforms. The software development kits (SDKs) provided by the platform proprietors (the Microsofts, Nintendos and Apples) establish the framework within which game development occurs. At stake here is the point Zittrain is making through his critique—platforms are not neutral. They bring with them technical constraints as well as commercial and legal provisions setting the terms and conditions governing access and use. Nick Montfort and Ian Bogost recognize the significance of platforms for digital media studies by launching what they call "Platform Studies" (also a book series with MIT Press) with the publication of their *Racing the Beam: The Atari Video Computer System*.[12] Montfort and Bogost maintain that the phrase "platform studies" does not necessarily imply a technological determinism in which the technical hardware of the platform directly affects particular outcomes.[13] They clarify, for example, that software is fundamental to the operation of platforms. In the case of the iPhone this is apparent with the significance of Apple's iOS and the

SDK. Furthermore, the SDK software environment developers use to make games for the iPhone is in some sense separate from the device and can be approached as a platform in and of itself.[14] Montfort and Bogost emphasize that with platform studies they seek to explore the connections and relations between technical specifics and culture: "Platform studies investigate the relationships between the hardware and software design of computing systems (platforms) and the creative works produced on those systems, which include but are not limited to video games."[15] These relationships are not only about the content of videogames but also the processes and cultural work of making videogames. Platforms are environments that enable and constrain certain ways of making and developing. They provide environments in and through which developers work.

In the process of building games for the iPhone platform and searching for new market opportunities for mobile devices, HalfBrick's developers are doing far more than designing and innovating new products: along the way, they are redesigning their firm and transforming their dominant mode of production—the project model of game development. They are reconfiguring their identities as producers, designers, programmers, marketing managers and CEOs. I call this phenomenon their "professional imaginary."[16] As David Stark comments in his ethnography of a new media startup, firms that pursue innovation in the context of unpredictable and uncertain environments "build organizations that are not only capable of learning but also capable of suspending accepted knowledge and established procedures to redraw cognitive categories and reconfigure relational boundaries—both at the level of the products and services produced by the firm and at the level of working practices and production processes within the firm."[17] Approaching the iPhone as a platform for innovation requires us to ask what kinds of innovation and what is at stake in these innovations. This is about much more than technical affordances or design qualities as it goes to the very heart of professional developers' working lives and environments and how they imagine what it means to be professional cultural producers.

INNOVATION AND VIDEOGAMES

Innovation can be a quite clumsy term: when expected to do too much rhetorical work, it can quickly become rather useless as an analytic category. As Brian Arthur comments, we still have "no deep understanding of what 'innovation' consists of."[18] My foundational definition of *innovation* borrows extensively from Arthur's recent *The Nature of Technology*. In the context of pursuing a more rigorous understanding of technology, Arthur contends that if we can understand the dynamics of technological change as evolutionary adaptation then we might also be able to better understand the processes of innovation.[19] Technologies, for Arthur, are combinations of existing technologies. This emphasis on combination and recombination draws on the Austrian economist Joseph Schumpeter's theory of innovation and economic change.[20] But, Arthur argues, this dynamic of recombination has a specific mechanism; it is evolutionary in that technological novelty emerges from what preceded it through something like heredity.[21] Arthur calls this emergence of technological novelty *"combinatorial evolution."*[22] As Lester and Piore observe, the mobile phone emerged from precisely such combinatory logics in "the space created by the ambiguity about whether the product was a radio or a telephone; by playing with that ambiguity, the device became something that was different from either of them."[23]

Arthur uses the term *innovation* to mean "novelty in technology," departing from Schumpeter's somewhat more constrained requirement that the novelty must also be put to commercial use. I adopt Arthur's[24] straightforward definition in this chapter and also follow his suggestion that it is important to ask questions about the mechanisms and dynamics of innovation through which novelty emerges. We need to ask these questions if we want not only to understand the novel outcome or product (the iPhone as device or the videogame as product) but also to describe and to explain the process and mechanisms through which these novel technologies, devices and products arise. This process is in many ways the dimension of *cultural* production when understood not only as cultural products but also in the dynamic sense as the processes and practices and identities through which culture is made and in turn makes us. But this dimension of identity goes missing from Arthur's definition and from his broader study of technology. Human agency is strangely absent. However, it is front and center in Schumpeter's work in his hero figure of the entrepreneur. In Schumpeter's account innovation invariably goes hand-in-hand with the identity and role of the entrepreneur. And, as we shall see, it is a certain mode of the enterprising developer that emerges from this account of HalfBrick.

Working with Arthur's starting definition of innovation, I now want to apply it in the context of videogames for the iPhone. The iPhone is an innovation in itself. But what do I mean when suggesting that videogames are in some sense innovative. What aspects or features of an iPhone game can be characterized as innovative? Furthermore, is it just the game design, for example, a particular game mechanic that uses the potential of the touchscreen, that is innovative, or are the processes of making, exploring and distributing that game also innovative? Or is the iPhone device innovative and the game title simply an expression or outcome of that innovation? As Jonathan Gray observes in the case of television entertainment, to some extent the very structures and practices of the industry can constrain creative innovation.[25] Both television and videogames are high-risk entertainment industries. Failure is endemic. And both are characterized by fundamental uncertainty in predicting what will be a success. The strategies both industries adopt to manage and reduce risk—formulaic standardization with repetition and imitation of the recent successful product category or genre—would seem to be inimical to innovation. Gray writes of television, and this most certainly also applies to videogames, "novelty and innovation, by executive logic, are risky, and while they may pay off big (a gamble that executives will sometimes take), it is safer, more 'comfortable,' to produce more of the same."[26]

However, as Ian Miles and Lawrence Green argue in the NESTA 2008 report, *Hidden Innovation in the Creative Industries*, much of the innovation processes and sources that characterize creative industries firms, such as videogames developers, are not captured by traditional innovation R&D indicators that dominate the manufacturing and high-tech sectors.[27] In this sense, these sources and processes of innovation are "hidden." They identify areas such as innovations in organizational forms and business models, including co-production approaches that integrate consumers in the product and experience development process.[28] Following Paul Stoneman's[29] work, Miles and Green also argue that the aesthetic, content and experience innovations that characterize the creative industries can be approached as forms of "soft innovation."[30] Hasan Bakhshi and Juan Mateos-Garcia observe, however, that the UK videogames industry, described as "one of the UK's unsung great economic success stories," is worryingly starting to struggle to innovate.[31] In 2009 they note that the UK industry started to shed staff, estimated at a 4 percent reduction in workforce, with 15 percent of the developer companies closing their doors. They argue that rapidly escalating development costs and a perceived skills shortage contribute to this worrying trend, while the preferred UK games development business model of relying on publisher-funded development (essentially a fee-for-service model in which the developer is unlikely to generate or retain original intellectual property [IP]) significantly contributes to this "innovation deficit."[32]

In 2010 the impact of the GFC on the global videogames industry also hit Australian developers hard. Rising development costs, combined with a strong Australian dollar, and a fee-for-service business model much like that in the UK, contributed to many developers struggling to adapt for a rapidly changing environment. Many of the more successful developers such as Brisbane located Krome Studios (Australia's then largest developer) had established international reputations of excellence for delivering fee-for-service games projects for the large US publishers. This meant Krome, like many other Australian developers, was not generating original IP or developing the skills for making original IP. This is not to suggest that there are not significant creativity or games design and development skills involved in making such games and growing a company that thrived in the competitive environment of the international videogames industry. It is arguable that Australian developers like Krome had found and locked-in to a local optimal solution for surviving the uncertainties and risks characterizing the videogames industry and that this required innovative skill. Nevertheless, in 2010, struggling to develop a successful original title and confronting a rapidly shifting and uncertain environment, Krome closed offices in Melbourne and Adelaide and shed staff from its Brisbane office. By October 2010 it was effectively out of business.

Over this period, however, other Australian developers adapted and reconfigured for the opportunity emerging in mobile games, especially by targeting the iPhone. Developers such as Melbourne's Firemint (established in 1999) and Brisbane's HalfBrick (established in 2001) reinvented themselves to pursue the opportunities provided by the growing market for casual games on mobile devices. Released on Apple's App Store on 5 March 2009, Firemint's *FlightControl*, an air traffic control simulation game, reached one million copies sold within three months and by January 2010 hit two million download sales. It continued to sell well throughout 2010 and made three million on the App Store by September. Firemint also enjoyed success on the iPhone and the iPad, with its car racing games *Real Racing*, and most recently with *Real Racing 2 HD* for the iPad. In May 2011 Electronic Arts (EA) and Firemint announced that EA had purchased Firemint.[33]

HALFBRICK'S *FRUIT NINJA*

HalfBrick built a reputation for developing solid licensed titles for platforms such as the Game Boy Advance, Nintendo DS and Sony's PSP. Recent stunning success with *Fruit Ninja* and other titles for the mobile games market has seen the company grow to some fifty staff. Released in April 2010 for iPhone and iPod Touch, in the first month *Fruit Ninja* reached over 200,000 copies sold and by the third month over one million download units sold. By September 2010 sales reached three million downloads, and by August 2011, these downloads had grown to over 36 million with close to 12 million paid downloads across all platform—iOS as well as Android and Windows Phone 7 systems.[34] In July 2010 *Fruit Ninja HD* was released for the iPad.

In discussions and semi-structured interviews with Shaniel Deo in 2010 and early 2011 that inform this research, Deo often referred to the iPhone as a platform that provides HalfBrick with opportunities to develop original IP and to explore a business and game development model that enables the company to thrive at a time when many other developers, both locally and internationally, have struggled. These discussions are peppered with his references to the iPhone as a gaming and development platform that enables HalfBrick to "self-publish." However, rather than emphasizing particular design features or game mechanics that contribute to the success of *Fruit Ninja*, Deo foregrounds the project processes of making videogames and the changing roles of developers. The following are the main issues he raises about developing for the iPhone: first, access to the growing casual and social gamer market, and second, significantly lower development costs and shorter development cycles when compared to making games for the major consoles. He mentions development cycles that are weeks or months rather than years, requiring significantly smaller development teams. This was often referred to as lowering barriers to entry and participation for a company such as HalfBrick. Deo suggests this change allows HalfBrick more latitude and opportunity to explore and experiment through trial-and-error rapid prototyping. As he put it,

> We can just try stuff out much more and see what happens, what works out. The guys can come up with and test new game mechanics. And we can afford to fail. That is huge for us, a big change. If we get it wrong, well we can learn from that, take what we think did work, maybe a game mechanic and then try again, tweaking and adjusting. Maybe we have lost a week or so of dev time. But it hasn't cost us the farm. We can try again until we get it right. It also means we can have multiple projects, multiple teams having a go at once.

Deo also suggests this allows the developer to be somewhat more responsive to feedback from the gamers. Deo calls this form of rapid prototyping game development "HalfBrick Fridays." I will return to how HalfBrick Fridays works as a search heuristic that enables HalfBrick to adapt for a rapidly changing environment. Third, he discusses the technical and design affordances of the iPhone and iOS. The touchscreen interface provides his developers and designers with an opportunity to find something new, a "new edge or new game mechanic." In the specific case of *Fruit Ninja*, which I describe in more detail in a moment, we see a compelling game mechanic that takes advantage of the haptic interface. You use your fingers to slice fruit. He also notes that Apple's development kit is relatively stable and robust to develop for. Fourth, Deo emphasizes the opportunity of digital distribution through the App Store. He describes this as enabling HalfBrick to become a "self-publisher." On this topic, he also notes the IP implications in that HalfBrick is now generating original IP. In an interview for creativeinnovation.net.au, Deo comments,

> The proliferation of downloadable platforms has really helped us—it means we can create and publish games ourselves without any middlemen. It's fairly straightforward becoming a registered developer for Apple—it's actually much harder to convince games publishers to produce your games.[35]

Fifth and finally, Deo emphasizes the marketing challenge of gaining attention for HalfBrick titles and generating the momentum needed to break into the App Store sales charts. He describes this variously as "going viral," "getting and keeping momentum" and "the importance of word of mouth." This engagement with social networking is increasingly integral to HalfBrick's games. Here he mentions the importance of Twitter, Facebook and YouTube. He also discusses the importance of maintaining relationships with online game press and timing HalfBrick's release of information to these press sources so as to maintain momentum towards securing a strong ranking on the App Store sales charts.

All of these factors contribute to how the iPhone functions as an innovation opportunity and platform for HalfBrick. In a moment, I will consider in more detail how HalfBrick organized, reorganized and adapted for this emerging opportunity. This quality of *emergent adaptability* is central to how Half-Brick innovates and to changes in the identities of professional developers.

THE iPHONE AS INNOVATION *PLATFORM?*

This idea of the iPhone as platform has been an organizing motif in my discussions with HalfBrick's Shainiel Deo and features heavily in the many interviews he and his staff give to games industry press and commentators. In "The politics of 'platforms'," Tarleton Gillespie examines and interrogates the discursive work the term "platform" does for content platform intermediaries such as YouTube as they position themselves for a quite diverse group of actors and associated interests including users, clients, advertisers and policy makers.[36] Gillespie describes how the term is deployed in a range of registers including populist appeals, technical affordances and sometimes as "platforms of opportunity." All these registers are evident even at this relatively early stage of research with Half-Brick; for example, in Deo's assessment of HalfBrick 's transformation as a "self-publisher." As in the case of YouTube, which is Gillespie's focus, the discourse of platform seeks to stitch together the diverse interests of advertisers, app developers and consumers. Furthermore, as Jean Burgess covers so well in her contribution to this book, all this is also shaped, although not determined, by the history of Apple's design and brand around values and discourses of usability and accessibility. Gillespie argues that this discursive work of platform functions to "carefully elide" the tensions and contradictions involved in serving and aligning these very different interests.

Is Apple's controlling role as intermediary elided through "platform's" discursive work? We could call this performativity "platforming." Gillespie[37] persuasively and carefully argues that this discursive work constructs a particular form or expression of "cultural imaginary" that allows providers,

> To make a broadly progressive sales pitch while also eliding the tensions inherent in their service: between user-generated and commercially-produced content, between cultivating community and serving up advertising, between intervening in the delivery of content and remaining neutral.[38]

He then suggests that this discursive work "arguably misrepresents the way YouTube and other intermediaries really shape public discourse online."[39] In the case of Apple and the iPhone, this discursive work is somewhat different as the values and affordances of openness and usability are very different. It would be a worthwhile exercise to examine in detail the various discursive registers through which iPhone as platform is performed, but this is beyond the ambit of this chapter. Gillespie is correct when he observes

> The business of being a cultural intermediary is a complex and fragile one, oriented as it is [in the case of YouTube] to at least three constituencies: end users, advertisers and professional content producers. This is where the discursive work is most vital. Intermediaries like YouTube must present themselves strategically to each of these audiences, carve out a role and a set of expectations that is acceptable to each and also serves their own financial interests, while resolving or at least eliding the contradictions between them.[40]

Gillespie adds that this is where the "real value" of the term "platform" is most apparent as it "brings these discourses into alignment without them unsettling each other."[41] This imaginary appeal to "'platform of opportunity'" is quite compelling in the case of games developers such as HalfBrick. Characterizing this as imaginary does not mean it is false. It certainly worked in the case of HalfBrick, as seen with the success of titles such as *Fruit Ninja*. But does this "platforming" work by eliding important tensions?

Gillespie concludes that terms such as *platform* "matter as much for what they hide as for what they reveal."[42] I agree that the YouTube business seeks to coordinate diverse interests and agents, often quite uncertainly and clumsily.[43] I also agree with Gillespie that it is important to foreground these complicated, tense and contradictory relationships. But do the logic and dynamics that articulate platforms such as YouTube and Apple's iPhone as innovation opportunities *necessarily* elide, obscure or resolve these tensions? These tensions and differences are very apparent and hardly hidden from view. They are right there on the surface of the platforms as it were and it does not take much digging for Gillespie to spot them. I doubt it is just the informed gaze of the academic researcher that can identify such dynamics. While making games for the iPhone HalfBrick's innovative opportunities were made possible *because* the developers searched for and maintained tensions, uncertainties and ambiguity concerning the iPhone's status as a platform. In the process, they reconfigure and indeed innovate the project form, which they call "HalfBrick Fridays," and the firm boundaries of making for such platforms. In all of this, perhaps we *see also* a rather different professional cultural imaginary forming, one that is less about eliding tensions and more about putting such uncertainties and frictions to work.

THE "DISSONANCE" OF HALFBRICK FRIDAYS—INNOVATING AS "HETERARCHY"

Economic sociologist, David Stark's recent *The Sense of Dissonance: Accounts of Worth in Economic Life* provides an important and helpful framework for rethinking and testing our assumptions about how

HalfBrick's developers' innovate in conditions of quite profound uncertainty.[44] His approach also resonates well with Arthur's[45] theory of how technology evolves through combinatorial dynamics. The analytic insight that emerges from Stark's comparative ethnographic research with four very different companies, including a new media start up, is that firms confronting evolutionary contexts of dynamic change may be best served by allowing, and indeed searching for and encouraging, multiple logics of worth and not by rushing to settle and resolve the resulting frictions. He calls the organizational forms that adapt practices of organizational diversity by harnessing the benefits and opportunities of such "dissonance" "heterarchies," which he defines as "an organizational form of distributed intelligence in which units are laterally accountable according to diverse principles of evaluation."[46] Stark suggests that agents in such firms were "benefiting from not asserting or fixing singular orders of worth and evaluation but maintaining an ongoing ambiguity among the co-existing principles."[47] Innovation and entrepreneurship then "is the ability to keep multiple evaluative principles in play and to exploit the resulting friction of their interplay."[48]

In Stark's framework of dissonance, logics of search come to the fore as organizing dynamics.[49] But he emphasizes a kind of search in which "you do not know what you are looking for but will recognize it when you find it."[50] The innovation problem and process for Stark thus concerns trying to recognize what is not yet formulated as a category and to make new connections and associations. The "word of mouth" social network market[51] dynamics that Deo and HalfBrick's director of marketing describe are precisely about this shift to search under conditions of uncertainty. As potential app purchasers, gamers search for which game should be on their iPhone. The social network market dynamics characterizing how an app rises to a prominent position on Apple's store sales charts suggests a new search strategy that cannot be quite encompassed by standard marketing classifications of demographic categories and associated social classifications of taste. The network, the social relations, is the source of value, perhaps even more so than the products.[52] These network relations increasingly define HalfBrick's brand.

But the more important point for the purposes of this chapter is how such search heuristics organize diversity.[53] In an evolutionary and complex adaptive system framework, this foregrounds adaptability at the organization level and the generative role of diversity.[54] Stark contends that,

> The adaptive potential of organizational diversity may be most fully realized when diverse evaluative principles coexist in an active rivalry within the enterprise. By rivalry, I refer not to competing camps and factions but to coexisting logics and frames of action. The organization of diversity is an active and sustained engagement in which there is more than one way to organize, label, interpret, and evaluate the same or similar activities. It increases the possibilities of long-term

adaptability by better search because the complexity that it promotes and the lack of simple coherence that it tolerates increase the diversity of options.[55]

This organizational diversity is evident in HalfBrick's internal project processes. "HalfBrick Fridays" commenced well before the *Fruit Ninja* success as a way of searching for opportunities other than the work-for-hire contracts for console projects and as a way to find time in the midst of the normal workload to explore smaller casual games. On HalfBrick Fridays, the developers brainstorm and pitch ideas in an iterative, rapid prototype process. Initially every two weeks and in the wake of *Fruit Ninja*'s success now weekly, they may come up with multiple contending and competing game ideas that quickly become playable prototypes—and then they evaluate those to find the one's they believe show something special by making it through the rounds of internal peer critique to essentially win an internal competition for attention, commitment, interest and resources. Diverse and competing understandings and principles of game design, social network dynamics and indeed of what the iPhone is and can be as both a gaming and social networking platform are all in play during these sessions and shape the following development project process. Indeed, I argue that the success HalfBrick enjoys from these Friday prototyping sessions relies on maintaining the tensions and frictions. Deo notes that between 5% and 10% of these prototypes go on to become a game. In describing this process of HalfBrick Fridays, he comments,

> We always liked the idea of retaining ownership of our games. But it was hard to dedicate our development teams to creating IP, because our commissioned projects took up a lot of time. So we introduced a system called "HalfBrick Fridays." On the first Friday, everyone would pitch their ideas to the rest of the company and whoever liked the idea could form a development team. Each team would devote the next four HalfBrick Fridays to building a prototype. At the end of each ten-week cycle, the teams would showcase their prototypes to the rest of the company. We used to have HalfBrick Fridays every fortnight, now we have them weekly.[56]

The successful projects that emerge from this process are then resourced for rapid development from prototype stage to an app ready to be released by upload to Apple's App Store. *Fruit Ninja* emerged through this process. In discussion Deo also notes that it is not just complete games that are identified through this process. He comments, "It may be a cool game mechanic that just works so well, but we don't quite know what to do with it, what to make of it. So it goes into the mix and we may well take it up and use it with a later prototype by bringing them together." HalfBrick's process of recombinatory innovation allows multiple ideas to emerge and develop; the developers may even borrow game mechanisms from each other's pitches, polishing and refining them through the titles that are eventually released.

In an interview for Kotaku, one of Australia's leading games industry commentary sites, Phil Larsen, HalfBrick's marketing director, notes that through this search, "We're instinctively trying to come up with the next idea for a big hit—it's hard! It's interesting—we can't always make the huge, phenomenal hit—no one can! You can't be super original every time."[57] He adds that in the case of *Fruit Ninja*, "We knew that was going to work when it was just a pitch, before it was even a game, we knew it would be awesome."[58] As Stark describes, this search heuristic cannot know in advance "exactly what it is looking for until it finds it."[59] The search challenge in an uncertain environment requires HalfBrick to continually redraw and reexamine internal boundaries and reinvent itself. And in the process, I suggest, they reimagine the role and work of cultural producers such as games developers. HalfBrick Fridays are not just about finding compelling games designs. These internal "social-network markets"[60] (and to what extent they are indeed "internal" and where the boundaries of these events are still needs to examined and described) are just as much about allocating creative work, resources and roles. It involves who is identified as "designer" by winning the support of colleagues to get the resources (time and other developers) to take the prototype to the next stage. Deo describes this as both "bottom up" and "creatively empowering." As do many of HalfBrick's developers. He also regularly raises the HalfBrick "profit shares" with the developers who work on the successful games.

When confronted by rapid technological change and transforming products and markets, such as with games and mobile devices, it is unlikely that there is one best solution or business model or project development approach or marketing strategy. And locking in to one solution may eventually erode the competitive advantage and market superiority that is initially secured. This "enterprising developer" model that HalfBrick experiments with is also then an emerging kind of entrepreneurial professional imaginary that contributes to shaping this search process. However, by entrepreneurial I do not mean reduced to profit incentive or to a narrowly conceived commercial identity. When Deo refers to "sharing profits," he never implies this is the primary motivation or only incentive driving the developers in the HalfBrick Fridays. At the same time, this is about the intrinsic motivation of crafting videogames and sharing the opportunity to do this. It is also about a firm coming up with an approach that keeps them in the game of a competitive and profoundly uncertain international videogames industry. It is a search heuristic that finds ways to skillfully adapt. It is not the creative and craft values of making innovative videogames opposed to the commercial values of profit. It is about harnessing the tensions and frictions between these various values. And negotiating such tensions is, I suggest, about identity formation.

This heterarchical organizational form that I suspect HalfBrick is adopting and adapting involves keeping its options open and continually experimenting with and redefining what constitutes an option, how to pursue that option and how to make and craft to take advantage of it. Stark comments,

> The problem for firms in uncertain environments is that the very mechanisms that foster allocative efficiency might eventually lock development into a path that is inefficient, viewed dynamically. Within this framework, our attention turns from a preoccupation with adaptation to a concern with adaptability, shifting from the problem of how to improve the immediate 'fit' with the environment to the problem of how to shape organizational structure to enhance its ability to respond to unpredictable future changes in the environment.[61]

The lesson here is that firms such as HalfBrick may sacrifice immediate allocative efficiency (adaptation) for dynamic efficiency (adaptability).[62]

HalfBrick Fridays introduces organizational diversity to provide the firm and its developers with the capacity to adapt when the environment changed. And this "heterarchy" expresses its evolutionary potential when different organizational principles "coexist in an active rivalry within the firm."[63] Innovative action in relation to platforms such as Apple's iPhone requires searching for and maintaining the tensions among contending and even formally incommensurable evaluative principles. Resolving or effacing such tensions may well lock firms into an innovation deficit situation.

CONCLUSION

The idea of Apple's iPhone as "innovation platform" looks somewhat shaky and unsettled in the wake of this encounter with HalfBrick Fridays and Stark's ideas of heterarchy, search, dissonance and adapting through organizational diversity, but for reasons somewhat different from those identified by Gillespie. Most certainly, there is innovative activity and the iPhone contributes to this, but is it an innovation "platform?" The problem, as Stark observes, is that the boundaries of the unit of action, the unit of innovation and the unit of entrepreneurship are not those of the legally bounded firm, or I would add, in this case, the iPhone device. It is more "distributed networks that span organizational boundaries."[64] In this chapter, I have barely scratched the surface of the various dissonances that shape HalfBrick's practices in developing videogames for the iPhone and other mobile devices. There is still much work to be done with opening the black box of organizational practices such as HalfBrick Fridays, let alone also investigating how such firms organize diversity with interactions through online social network "platforms" such as YouTube, Twitter, Facebook and so on. We need to better understand and explain the specific mechanisms and dynamics of these dissonant practices that may well increasingly be at the core of successful innovation practice and of professional creative producers' identities. Moreover, how does the materiality and affordances[65] of devices such as iPhones and iPads contribute to how we search for and connect with dissonant opportunities? As intimated throughout this chapter, I suggest this should be approached as an evolutionary dynamic in the context of complex, adaptive systems.[66] In all of this, however, and I think evident throughout this account of HalfBrick, are questions about the professional skills needed to craft and make and operate in the context of dissonant innovation. And here, I glimpse a certain productive dissonance with Gillespie's work I discussed earlier. Just as the discourse of "platform" performs a certain cultural imaginary, what are the implications of a "professional imaginary" performed by adapting for and through dissonance?

ACKNOWLEDGMENTS

The idea of "professional imaginary" that I start to explore in this chapter first arose through discussions with Fred Turner and Seth Lewis at a National Science Foundation supported workshop at Cornell University, Department of Communication (19–21 March 2011) on the topic of examining contemporary culture digital production. In fact I believe that it was Turner who first coined the term "professional imaginary" at the workshop. Many of the ideas in this paper started to take form through the discussion and dialogue at this workshop, especially with Turner, Lewis and Thomas Malaby. These discussions are continuing through a workshop blog (culturedigitally.org). I would also like to thank my colleagues Jason Potts and Stuart Cunningham who collaborate with me on the research project that grounds this chapter. Thanks to Jean Burgess for her thoughtful and incisive commentary on earlier versions of this chapter. Thanks also to Justin Brow for helping me to develop the relationship with HalfBrick. Finally, thanks to Shainiel Deo (CEO of HalfBrick) for his generosity in discussing these topics and providing access to his company.

NOTES

1. Lev Grossman, "Invention of the year: the iPhone," *Time.com* (1 November 2007), http://www.time.com/time/specials/2007/article/0,28804,1677329_1 678542_1677891,00.html (accessed 15 July 2011).
2. Ibid.
3. This project is an Australian Research Council Linkage grant project (LP100200056), "The games and the wider interactive entertainment industry in Australia: an inquiry into sources of innovation." Researchers participating in this project include Stuart Cunningham, Jason Potts and Karen Pearlman. The Australia Council for the Arts also supports the project.

4. Tim Merel, "The Big V: the great games market split," *Venturebeat.com* (6 July 2011), venturebeat.com/2011/07/06/the-big-v-the-great-games-market-split/?obref=obinsite (accessed 15 July 2011).

5. Nielsen, "play before work: games most popular mobile app category in US," *Nielsenwire.com* (6 July 2011), blog.nielsen.com/nielsenwire/?p=28273 (accessed 15 July 2011).

6. Ibid.

7. Jonathan Zittrain, *The Future of the Internet—And How to Stop It* (New Haven & London: Yale University Press, 2008), 3.

8. Ibid.

9. Ibid.

10. Gerard Goggin, "Adapting the mobile phone: the iPhone and its consumption," *Continuum* 23(2), 2009: 231–244.

11. Mark Andreessen, "Analyzing the Facebook platform, three weeks in," blog posted on 12 October 2009, http://pmarca-archive.posterous.com/analyzing-the-facebook-platform-three-weeks-i (accessed 23 July 2011).

12. Nick Montfort and Ian Bogost, *Racing the Beam: The Atari Video Computer System* (Cambridge, MA: The MIT Press, 2009).

13. Ian Bogost and Nick Montfort, "Platform studies: frequently questioned answers," *Digital Arts and Culture Conference*, 2009, http://www.bogost.com/downloads/bogost_montfort_dac_2009.pdf (accessed 23 July 2011).

14. Ibid.

15. Ibid.

16. This idea first emerged through discussions with Fred Turner at a National Science Foundation supported workshop at Cornell University, Department of Communication (19–21 March 2011). Turner first coined this term in these discussions.

17. David Stark, *The Sense of Dissonance: Accounts of Worth in Economic Life* (Princeton, NJ: Princeton University Press, 2009), 83.

18. W.B. Arthur, *The Nature of Technology: What It Is and How It Evolves* (New York: Free Press, 2009), 13.

19. Ibid., 15.

20. Joseph A. Schumpeter, *The Theory of Economic Development* (Cambridge, MA: Harvard University Press, 1934).

21. Arthur, op cit., 18–19.

22. Ibid., 22.

23. Richard K. Lester and Michael J. Piore, *Innovation: The Missing Dimension* (Cambridge, MA & London: Harvard University Press, 2004), 181.

24. Arthur, op cit., 90–91.

25. Jonathan Gray, *Television Entertainment* (New York & London: Routledge, 2008), 23–24.

26. Ibid., 24.

27. Ian Miles and Lawrence Green, *Hidden Innovation in the Creative Industries: Research Report* (National Endowment for Science, Technology and the Arts, London, UK, 2008), www.nesta.org.uk/publications/reports/assets/features/hidden_innovation (accessed 15 July 2011).

28. Ibid., 13; *see also* 25–27.

29. Paul Stoneman and Hasan Bakhsi, *Soft Innovation: Towards a More Complete Picture of Innovative Change* (London: National Endowment for Science, Technology and the Arts [NESTA] 2009): www.nesta.org.uk/library/documents/Report%2022%20%20Soft%20Innovation%20v9.pdf (accessed 15 July 2011).

30. Miles and Green, op cit., 14.

31. Hasan Bakhshi and Juan Mateos-Garcia, *The innovation game: adjusting the R&D tax credit: boosting innovation in the UK video games industry*

(London: NESTA, 2010), www.nesta.org.uk/library/documents/The_Innovation_Game_-_FINAL.pdf (accessed 15 July 2011).

32. Ibid., 6.
33. Asher Moses, "Newest Aussie high-tech tycoon's multimillion-dollar deal," *Theage.com.au*, 4 May 2011, theage.com.au/digital-life/smartphone-apps/newest-aussie-hightech-tycoons-multimilliondollar-deal-20110504–1e7sk.html (accessed 15 July 2011).
34. E-mail from Phil Larsen, HalfBrick's marketing director.
35. Shainiel Deo, "Interview with Creativeinnovation.net.au—commercialising IP: HalfBrick" *Creativeinnovation.net.au*, 2010, creativeinnovation.net.au/Features/business-management/commercialising-ip-halfbrick.html (accessed 15 July 2011).
36. Tarleton Gillespie, "The politics of 'platforms'," *New Media & Society* 12(3), 2010: 347–364; Gerard Goggin, "Adapting the mobile phone: the iPhone and its consumption," *Continuum* 23(2), 2009: 231–244.
37. Ibid., 348.
38. Ibid.
39. Ibid., 349.
40. Ibid., 353; *see also* 358.
41. Ibid., 353.
42. Ibid., 359.
43. Jean Burgess and Joshua Green, *YouTube: Online Video and Participatory Culture* (Cambridge: Polity Press, 2009).
44. Stark, *Sense of Dissonance*.
45. Arthur, *Nature of Technology*.
46. Stark, *Sense of Dissonance*, 14.
47. Ibid., xxiv.
48. Ibid., 15.
49. Ibid., 169–170.
50. Ibid., 2–6.
51. Jason Potts, Stuart Cunningham, John Hartley and Paul Ormerod, "Social network markets: a new definition of creative industries," *Journal of Cultural Economics* 32, 2008: 167–185; John Banks and Jason Potts, "Co-creating games: a co-evolutionary analysis," *New Media and Society* 12(2), 2010: 253–270.
52. Stark, *Sense of Dissonance*, 173.
53. Ibid., 164.
54. Scott E. Page, *The Difference: How the Power of Diversity Creates Better Groups, Firms, Schools, and Societies*, (Princeton, NJ: Princeton University Press, 2007); Scott E. Page, *Diversity and Complexity* (Princeton, NJ: Princeton University Press, 2011).
55. Ibid., 26–27; *see also* 164.
56. Deo, "Interview."
57. Mark Serrels, "HalfBrick and Firemint: making lightning strike twice," Kotaku.com.au, 13 April 2011, www.kotaku.com.au/2011/04/442189/ (accessed 15 July 2011).
58. Ibid.
59. Stark, *Sense of Dissonance*, 174.
60. Potts et al., "Social network markets."
61. Stark, *Sense of Dissonance*, 178.
62. Ibid., 178.
63. Ibid., 179.
64. Ibid., 196.
65. "Affordances" is an unfortunately clumsy term, and I think Arthur's (2009) work on technology may provide us with a framework for improving our understanding of the relationships between technologies and innovation.

66. I have elaborated on this elsewhere with my colleague Jason Potts (Banks, "Co-creating games"). The recent excellent popular work *Adapt: Why Success Always Starts with Failure*, by economist Tim Harford (London: Little Brown, 2011), is also currently generating some interest in how evolutionary principles of adaptive, trial-and-error processes may contribute to a better understanding of fostering and generating innovation.

11 Network Labor
Beyond the Shadow of Foxconn

Jack Linchuan Qiu

INTRODUCTION

Five days after I attended the iPhone Workshop held at Queensland University of Technology (QUT) on 11 July 2009, a 25-year-old worker named Sun Danyong committed suicide because an iPhone 4 prototype was missing in his factory, Foxconn. The security guards interrogated him. He was beaten, threatened and locked up. At 1:48 a.m., he sent his last SMS to his girlfriend:

> Dear, I'm sorry. Go back home tomorrow. I ran into some problems. Don't tell my family. Don't contact me. I'm begging you for the first time. Please do it! I'm sorry.

He then jumped from the twelfth floor of an apartment building.[1]

At the QUT Workshop, I shared my observations about work injuries in Foxconn, the manufacturer that makes all Apple products. In international press, the Longhua campus of Foxconn in Shenzhen, Guangdong Province, had already been known as "iPod City" due to its enormous scale and harsh conditions exposed by a British newspaper in 2006.[2] I had also been visiting hospitals filled with injured Foxconn workers, who typically lost their fingertips producing mobile phones and computers.

I thought these injuries were the cruelest reminder about the material dimension of this "weightless" "fingertip" economy that the iPhone is too often believed to represent, and central to this material dimension is the indispensable factor of human labor in electronics manufacturing, a factor too often missing in studies of digital culture. Were fingertip injuries the cruelest reminder? Sun Dandong showed that I was wrong. He was only the beginning. From January to August 2010, at least seventeen Foxconn workers reportedly committed suicide by jumping from tall buildings. Only four survived, two with permanent disabilities.[3] Never was there such a tidal wave of worker suicides in the history of electronics manufacturing, anywhere in the world. What went wrong? How could the making of the iPhone—a device seen as fun, cool and harmless—become so fatal? These questions are examined in this chapter via the concept of network labor.

This chapter first introduces the notion of network labor before a more systematic examination of China's contemporary labor formations, seen through the prism of Foxconn, along with the associated new media politics, especially in the aftermath of the suicide tragedies. What happened in Foxconn was the symptom of an unsustainable model of global labor. It calls for a fundamental rethinking of the role of material labor as the basis for a new media culture signified by the iPhone.

NETWORK LABOR

According to Manuel Castells, the network is a fundamental form of social organization.[4] Much of modern history is, however, characterized by the competition between two other forms of social organization: statism and capitalism. Since the end of the Cold War, the network has become more prevalent and predominant, a trend precipitated first by the development of the internet since the 1990s, then by the diffusion of mobile communication and the "mobile network society" since the turn of the century.[5]

The network logic has been spreading.[6] On the one hand, it erodes the notion of exclusive sovereignty in statism, leading to a new power structure called the "network state" that can be seen as the organizing principle behind such supranational entities as the European Union, World Trade Organization and G20, to which China belongs as one of the key states. On the other hand, it interacts with market capitalism and dissolves traditional corporate boundaries, producing "network enterprise," which continues the trend of post-Fordism, Toyotism and, in the context of Southeast Asia, further expansion of Chinese business networks.[7] The network state and network enterprise are the two central institutional pillars within Castells' theorization of network society in the 1990s.

By *network labor*, I mean "a materializing pillar of the network society, parallel to the emergence of the network enterprise and the network state, globally and regionally."[8] The concept of network labor reflects several general themes in Castells' theory of network society but takes them closer to empirical reality a decade into the new century. First, it is about the further spread of network logic, from the domains of governments and corporations to the realm of labor. Second, labor has always been crucial to network society, in which the contradiction between high-end "self-programmable labor" and low-end "generic labor" was seen as the most definitive feature in post-industrial economies.[9] Third, network society builds on global connectivity, but its formation in the global "space of flows" is highly uneven, thus producing certain preferred city regions of concentrated network activity.[10] A major transformation of recent years is the rise of emerging economies, especially China and India, as centers of the global IT industry in terms of hardware or software production.[11] These new industrial centers in China and India are, unsurprisingly, also centers of network labor formation. Finally, the network society is not just about top-down control imposed by state agents or capital. It is also about grassroots social movements and the "power of identity" and of tradition, now enhanced by ICT-based mobilization networks.[12] In labor movements, this was best exemplified by the Honda strikes in Guangdong Province during May and June 2010, when factory workers used mobile phones and the internet to circulate not only logistical information for collective action but also photographs and videos shot on their mobile phones.[13]

What exactly is network labor when it comes to the analysis of iPhone manufacture and of Foxconn? The most fundamental dimension is certainly still wage labor, responsible for the material production of mobile phones and computer gadgets. Of Foxconn's more than one million employees in mainland China, most are involved in labor processes of material production that span from the design and tuning of machinery to the molding, polishing and assembly of parts to packaging and transportation. They include white-collar self-programmable labor as well as blue-collar generic labor. But the majority of them belong to what I term "programmable labor," which is an extension of Taylorism in the twenty-first century. Work procedure is simplified, "unskilled," calculated by computer to the precision of certain seconds per movement of the worker's arm, turning workers into nothing but "programmed" parts of the industrial machine. The daily quota of a female worker, for example, is to put 5,800 tiny screws onto 2,900 Mac SuperDrives. In so doing, she provides low-end network labor to Foxconn while becoming a low-end network laborer herself.

Inside Foxconn, Apple and their collaborators, there are other employees engaging in the more "immaterial" aspects of production. These are workers who design, install and test software, who provide content, applications, sales and post-sale services, as well as those working in Lazzarato's now classic categories of "immaterial labor": advertising, marketing and public relations.[14] These employees are better paid than workers on the assembly line. While some of them belong to the category of self-programmable labor, others still perform repetitive tasks—like software testing or quality control—that were pre-programmed for them. The line between material and immaterial labor is therefore fuzzy because both categories fall under the general logic of network enterprise. They are, in this sense, nothing but two variants of network labor.

The concept of network labor, however, encompasses not only waged labor. It extends to unwaged labor as well, resulting from the expansion of network enterprise into the "virtual" economy. One indicative phenomenon is that the internet cafes close to Foxconn are often packed at around 8 a.m. because, after coming off night shift, many male workers have the habit of playing online games, watching internet videos or chatting with friends online. Others with their own computers or internet-capable mobile phones spend more time online, entertaining themselves, killing time or seeking information, and thus creating value for internet companies or telecom corporations. This is similar to the trend of "precarious playbor" identified in the context of online gaming in Western countries, where dedicated gamers modify and improve the games they love, but the improvements, intellectual property rights and financial benefits all belong to the online game corporations.[15] Yes, the gamers may enjoy the process of playing online games and modifying them. So do Foxconn workers going into internet cafes or playing with their mobile phones. However, from a critical perspective, these factory workers can be seen as exploited just like the precarious playbor, as information have-less users joining the ranks of upper- and middle-class users who create value for the virtual economy by contributing attention to online content and, sometimes, producing their own user-created content (UCC) such as text messages, blogs and images.[16]

Network laborers are also "users" of iPhones, authentic or fake. Some of their mobile handsets carry brands like iPhoen or iPheno and cost around $60. Others spend much of their weekends looking into more high-end fake iPhones sold for $100 to $150; these look almost identical to authentic ones, complete with Wi-Fi and 3G connectivity, though they cannot download Apple applications. This is part of the "bandit phone" (*shanzhaiji*) phenomenon that I have discussed elsewhere; as a model of technological innovation, *shanzhaiji* can be seen as an alternative network formation that challenges the power of big brands as well as state authorities.[17] In the case of Foxconn workers as end-users trying to find the best fake iPhone, the immediate effect is further alienation beyond the workplace, into the realms of entertainment and sociability, where the myth of the iPhone is reproduced through an informal economy and everyday life activities.

However, besides this immediate effect, it is also important to recognize that the widespread use of working-class ICTs, including bandit iPhones, opens up opportunities for the building of translocal labor networks, a new class-consciousness and transnational solidarity.[18] Although the role of network labor as an agent of social change has only started to manifest itself, this role is probably the most important historical mission of network labor at a structural and global level.

INSIDE THE WORLD'S FACTORY

Before a focused examination of Foxconn, it is necessary to contextualize our discussion against the backdrop of China's rapid emergence as the world's factory. Consumer electronics has been a global industry since the spread of radio broadcasting in the first half of the twentieth century. During the post-war period, Japan emerged as the world's epicenter for the manufacture of consumer electronics, a pattern that later spread to other newly industrialized Asian economies: South Korea, Taiwan, Malaysia and

Thailand. For most of this global and regional history of electronics manufacturing, China was absent primarily due to the Chinese Communist Party's (CCP) preference of heavy industry over consumer products from the 1950s well into the 1980s.

However, since the 1990s, China has become the world's factory for consumer electronics as it has for so many other things, from apparel to toys to stationery. The manufacture of consumer electronics has centered on two coastal regions, the Pearl River Delta in the South and the Yangtze River Delta in the East, which are both home to Foxconn's largest and most important production facilities today. Initially, plants were set up in these coastal regions to make less sophisticated products such as electronic watches, radios and pagers, when Foxconn was still a nondescript factory. But since the turn of the century, cameras, MP3s, mobile phones and desktop and laptop computers have become the fastest growing Chinese outputs.

This was the beginning of a new era for global capitalism, whose growth, according to Dan Schiller,[19] relies mostly on "two poles" of capital accumulation: one being the IT industry, the other China. This conceptual framework anticipates the emergence of China as a new center for global electronics manufacturing, including the rise of Foxconn on the back of Chinese workers. However, if examined more closely, this development has been anything but obvious. Its contours are full of tension and conflicts, meaning that the often circulated media narrative—China has the world's cheapest labor; therefore, it is "naturally" the world's factory—is incorrect.

As mentioned, China's role in the history of consumer electronics is relatively new. Although China has the world's largest population, its labor reserve was separated from the realm of global manufacture until the early 1990s, when Chinese authorities accelerated pro-capital reform with special policies to aid export-oriented manufacturing in the Pearl and Yangtze River Deltas. Without state intervention, the sheer size of the labor force is at best a necessary but not sufficient condition to make China the world's factory.

There are many results of state intervention; some are intended, like the restructuring of Chinese industrial structure away from heavy industry and the rapid growth of export-oriented coastal regional economies, and some are unintended, like massive lay-offs and labor unrest in old industrial zones of the northeastern regions. The combined consequence is clear: Chinese workers—now consisting mostly of young migrants from rural China—are forced to work even harder for less pay.

Reaching this goal of further exploitation is certainly not an easy process, considering that Chinese workers were already poorly paid in the pre-reform era. According to Larry Lang, an influential economist in Hong Kong, total wages in China have been declining as a percentage of the country's national GDP.[20] In 1978, all employees in China received 15.7 percent of national GDP as their total wages. The percentage was 15.5

percent in 1989 but dropped to 12.1 percent in 1999. In 2009, it was only 8 percent! This was in comparison with 58 percent in the US, 44 percent in Korea, 28 percent in Thailand and 24 percent in Iran.

Accompanying the tightening squeeze on Chinese labor is the rapid increase in labor disputes in the country. In the first decade of the twenty-first century, the total number of labor disputes in China grew at an annual rate of nineteen percent, culminating in 684,400 incidents in 2009.[21] Worker resistance has, in this sense, always been an integral part of contemporary labor politics in China. Its acceleration in the IT industry was well under way long before the suicide tragedies occurred in Foxconn in 2010. In March 2008, for example, a series of conflicts happened in an IBM subcontractor in Dongguan, Guangdong Province, which included four cases of worker suicides.[22]

The rise of China as the world's factory for consumer electronics is therefore only partially explained by the capitalist "race to the bottom" at the global level, which wouldn't have worked without the active participation of regional business networks under the wing of the Chinese authorities. The combined result is the shaping of a precarious labor force under the control of a new "high-tech flexible production" regime in China's IT manufacture sector that accounts for the fastest growing bulk of the country's exports.[23]

Year 2009 was also a special year as it immediately followed the 2008 global financial crisis. At the beginning of 2009, 20 million Chinese migrant workers lost their jobs, which was equivalent to 80 percent of the total job loss in all OECD (Organization for Economic Co-operation and Development) countries.[24] Behind the numbers were numerous dramatic events of worker suffering. For example, He Jinxi, a jewelry-factory worker who was laid off and beaten by security guards, stabbed two human resources managers to death and then took his own life in March 2009, leaving behind his widow who was seven months pregnant.[25] Liu Hanhuang lost his right hand while working in a Taiwanese-owned metal-part factory. In June 2009, he used his left hand to kill two Taiwanese managers because the factory owner refused to pay him the full compensation ordered by court. While families of the Taiwanese managers pled for mercy on his behalf, Liu asked the judge to give him the death sentence, which was granted.[26]

The year 2009 was indeed full of such incidents, including the death of Sun Danyong in July 2009. This was almost half a year before the wave of suicides began in Foxconn in January 2010. It was also in this year when newly employed Uyghur workers clashed with ethnic Han workers in the Xuri Toy Factory in Shaoguan, Guangdong Province, sparking ethnic riots thousands of miles away in Urumqi, in China's Muslim northwest. What started as labor disputes along the coast ended up as ethnic riots in the hinterland, showing again the powerful network effect of labor politics that may trigger other confrontations in the country.

Ironically, 2009 might also be remembered for *Time Magazine*'s honoring of "The Chinese worker" as its "Person of the Year Runner-Up," ranking higher than US President Barack Obama. Their credit: "leading the world to economic recovery" because their hard work allowed China's continued economic growth that proved to be "an economic stimulus for everyone else."[27] The workers selected by *Time* all came from Shenzhen Guangke Technology Co., an electronics factory located less than one hour from Foxconn.

THE RISE OF FOXCONN

The story of Foxconn began in 1974 when Taiwanese businessman Terry Kuo founded Hon Hai Plastics Corporation in Tucheng, Taiwan, to produce parts for black-and-white televisions.[28] In 1981, Hon Hai had moved into the new market of computer connectors and adopted Foxconn as its trade name for operations outside Taiwan. Before long, Terry Kuo gained a notorious reputation among Taiwanese electronics manufacturers for being parsimonious and extremely demanding. According to an industry insider, a popular saying about him at the time was "You want his money, he wants your life!"[29]

In 1988, Foxconn established its first factory in mainland China. In 1991, Hon Hai went public on the Taiwanese stock market. The public offering, support from Chinese authorities and the influx of migrant workers from rural China allowed Foxconn to expand its mainland facilities in leaps and bounds throughout the 1990s, enabling it to become the world's leading manufacturer of computer motherboards at the turn of century. By then, Foxconn had begun to make iMac computers for Apple while also producing components for leading brands in mobile phone and video game console markets. Rapid growth continued in the 2000s. In March 2005, *Forbes* identified Terry Kuo as the richest Taiwanese tycoon, a position he held for five years until the effect of the suicide tragedies relegated him to the second position in 2010.

Year 2006 marked a turning point for Foxconn. Before then, Terry Kuo had been running the giant corporation in a secretive manner, about which investors in the stock market often complained. Such complaints were, however, easily marginalized by news of Foxconn's further expansion and reports of its record-breaking profits. When less-than-favorable reports about the company appeared in Taiwanese press, Kuo would deploy his formidable legal team to tackle the journalists and news organizations heavy-handedly. One such incident led to a high-profile confrontation with the Association of Taiwan Journalists, which accused Kuo of threatening press freedom in November 2004,[30] although such confrontations were little known outside Taiwan.

In June 2006, a muckraking report about Foxconn appeared in the *Mail on Sunday*, a British newspaper, which exposed awful conditions in the company's largest plants in mainland China.[31] Describing the Longhua campus of Foxconn, the article reads, "It's a sprawling place where 200,000 people work and sleep—meaning this iPod City has a population bigger than Newcastle's." It went on to reveal harsh working conditions, wages as low as £27 per month and photographs of horrible living conditions in the dormitories.

By coining the phrase "iPod City," the report set off a serious public relations crisis for both Apple and Foxconn. The story was relayed in other international media channels including the BBC and *MacWorld* as well as Chinese media. Some Chinese citizen journalists sneaked into Foxconn and came back with even more damaging images and findings; for example, some workers were only paid 490 yuan per month, well below the minimum wage of 580 yuan per month in the region. Kuo was irritated and tried to silence the reports. But this time, it didn't work.

In July 2006, Foxconn sued *First Financial Daily (FFD)*, a Shanghai newspaper, rather than the *Mail on Sunday*, BBC or *MacWorld*. Foxconn claimed that reports by *FFD* caused it significant reputation damage, worth 30 million yuan (about US$5 million). Foxconn lawyers sued two *FFD* journalists and asked the court to freeze all their assets. The legal action triggered public uproar especially in online forums and among journalistic circles. But for more than a month, Foxconn showed no sign of backing down until Sina, an internet portal with deep connections to China's top leadership, highlighted the dispute on its website in August 2006.

With Sina joining the dispute and siding with its critics, Foxconn's stance suddenly softened. In less than 63 hours, it announced that the real purpose of its legal action was reputation not compensation, and that no matter how much it could win from the case, it would donate the full amount to charity. At 10 p.m. on the same day, Foxconn sent out a second news release stating that it would demand only 1 yuan instead of 30 million. Three days later, Foxconn completely backed down, not only dropping the case but releasing a joint statement with *FFD* saying that the two parties would work together to promote a "harmonious society" along the official line of the Chinese authorities.

In China, the 2006 dispute brought unprecedented attention to Foxconn, now known throughout the world as "iPod City." It was no longer possible for Foxconn to hide its sweatshop conditions from the rest of the world. Under public pressure, Foxconn began resolving some of its labor issues on the surface level by building better dormitories and recreation facilities and raising wages, while continuing its military-style management framework that atomizes workers.[32]

Meanwhile, probably not coincidentally, Terry Kuo launched an unprecedented media campaign from January 2007 to mid-2008 by transforming himself into a womanizer, completely different from his earlier image of a mean, tasteless, workaholic tycoon. Over this period, he was frequently rumored to have high-profile affairs with several famous actresses, singers and artists in both Hong Kong and Taiwan, allowing himself to be caught by gossip news reporters late in the night accompanying the rumored females, sometimes hand in hand. Entertainment magazines used these images on their front covers. The gossip news about his private life succeeded in distracting public attention, at least temporarily, from Foxconn's labor issues. The media profited from his media offensive as well by selling a large number of tabloids reports.

In August 2008, another dramatic incident occurred, when a few images were found in a newly bought iPhone in the UK, which happened to be where the "iPod City" story was first published. They showed a beautiful female worker in a white uniform, smiling and posing for the pictures at what appeared to be her work desk on an iPhone assembly line (Figure 11.1). Reporters found out that these were probably taken by Foxconn

Figure 11.1   The "iPhone girl" image discovered in August 2008.

workers in the quality control division testing the iPhone camera, although according to the production procedures, it was almost impossible for such images not to be deleted before they reached the customer.[33] Chinese netizens tried to find this female worker, now known throughout the world as the "iPhone girl." All they could find was her surname, Li, and the unconfirmed report that she was already fired for not abiding by production rules. To this day, the iPhone girl remains a mysterious part of the Foxconn story. No one knows whether the images were left in the phone unintentionally or whether they were another deliberate attempt on behalf of the company to divert public opinion.

Before the iPhone girl disappeared from news headlines, the global financial crisis began. Global demand for consumer electronics declined sharply after the fall of Lehman Brothers in September 2008. Foxconn responded almost immediately by laying off workers. In October 2008, it fired about 100,000 of its 500,000 workers in mainland China. In December 2008, another 40,000 were laid off. However, in March 2009, as China's exports started to recover, and as global iPhone sales accelerated, Foxconn added 30,000 workers almost overnight.[34]

"How could Foxconn hire and dismiss tens of thousands of workers so easily? What tricks does it have in playing with human capital so successfully?" wrote an article in *China Information World*.[35] It discusses the various ways in which Foxconn can maximize its profit by evading labor laws and its basic corporate social responsibilities. What this article fails to see is the human cost behind Foxconn's merciless "flexible" labor management regime, which led to the suicide of several Foxconn workers since June 2007, Sun Danyong's death in July 2009 and ultimately the series of suicide tragedies in the first half of 2010.

Never in the entire history of industrial production was there such a long stream of desperate workers taking their own lives. In a way, electronics manufacturing differs from many other industries—such as mining, chemistry, electric power generation and paper production—in that its work injuries or occupational diseases are not usually fatal under normal circumstances. Workers may lose their fingertips, but not their lives. However, as the specter of suicides hounded Foxconn in 2010, electronics manufacturing—carried out under the military-style factory regime—turned out to be just as deadly.

THE NETWORK BLOWBACK

Foxconn is among the latest epitomes of network enterprise, whose logic of flexible accumulation has shaped the life and work of its employees. This network extends from Foxconn and its suppliers in the Asian Pacific region to Apple, other major brands and their partners globally. In this sense, network labor is an inevitable consequence of network enterprise:

although ordinary Foxconn workers may be unaware of the network structures beyond the shop floor, their mode of production has been fundamentally affected by this global network of consumer electronics, in which manufacturing is merely a less glamorous link between high-tech research and development on the one hand and high-volume marketing and consumption on the other.

Although the network enterprise took a hit in the 2008 global economic crisis, by the summer of 2010, Foxconn had fully recovered. Benefiting from the worldwide iPhone and iPad craze, its expansion reached an unprecedented scale. The Longhua campus in Shenzhen, i.e., the "iPod City" reportedly accommodating 200,000 workers in mid-2006, now had about 400,000 workers producing not only iPhones and other Apple products but also Nokia, Sony, HP and Dell products. The series of suicides occurred mostly in this massive factory facility, but they only temporarily disrupted Foxconn's daily operations, in part because the network structure helped distribute risk both within the industrial system of global electronics manufacturing (e.g., Apple giving Foxconn additional shares of revenue to increase wages) and beyond (e.g., subcontracting dormitory management to local companies).

The network state poses another risk for Foxconn and, one may argue, an opportunity for the mobilization of network labor. As the wave of suicides intensified in late May, China's Ministry of Labor and Social Security, the Ministry of Health and All China Federation of Trade Unions (the official union) sent investigative teams into Foxconn. The results of their investigation are, however, mostly hidden from public view. On 26 May 2010, Foxconn was also open to the media, albeit only for a single day. Soon, in early June, Chinese authorities banned further coverage of Foxconn because news reports on suicides allegedly led to copycats. Until this day, the public remains uninformed about the particular reasons that caused each death, who was held accountable and what measures were taken to prevent future tragedies.

According to Chan and Pun, the Foxconn suicides should be seen as protests by desperate workers. "Their defiant deaths demand that society reflect upon the costs of a state-promoted development model that sacrifices dignity for corporate profit in the name of economic growth."[36] More specifically, they point out three components in Foxconn's problematic development model including (a) unethical purchasing practices of leading international brands like Apple, (b) abusive and illegal management methods within Foxconn and (c) local Chinese officials colluding with Foxconn and Apple in neglecting workers' rights. The network state, in this sense, is an active participant in the making of network labor, following the lead of network enterprise. However, in so doing, it also becomes as precarious as network enterprise once alternative networks, online and offline, are mobilized by the suicide tragedies.

In mainland China, the mass media system is under direct control of the network state. So the issues were marginalized, even completely silenced, when the window of reporting Foxconn tragedies closed in June 2010.

However, the network logic has infiltrated society, including labor. Almost all workers have mobile phones, and the great majority of them are internet users.[37] They were already using SMS, Weblogs and QQ (the most popular instant-messaging and social networking service among Chinese workers) in everyday life. Now that a new wave of labor-capital clashes was triggered by the suicides—helpless individual acts in the beginning but with powerful butterfly effects within and beyond Foxconn—the tools of everyday connectivity were converted, almost instantly, into tools of labor solidarity.

Strikes at the Honda factory, less than two hours away from the Foxconn facilities in Shenzhen, were organized by workers using low-end camera phones. As a Reuters photograph at the time shows, female Honda workers were all holding up their phones to capture images of the security guards as a way to not only record the scene but also deter brutal suppression by management or the local state.[38] Many of these images were indeed circulated online, which stimulated more industrial actions in electronic manufacturing plants in other parts of the country, for instance, in Dalian of northeast China.[39]

In addition to spontaneous networking among ordinary workers, a backbone of network labor advocacy is provided by grassroots labor NGOs, including their online and offline networks with regional and global reach. Some of these NGOs are affiliated with the authorities and run as mini-bureaucracies, others are relatively autonomous and run as loose, informal networks. Some use funding from official or commercial organizations in the country, while others depend on donations from citizens and charities abroad. Regardless of their operational models, in general they all engage in workers' rights advocacy, relying heavily on working-class ICTs, i.e., inexpensive internet and mobile phone services such as QQ. In so doing, they produce a particular kind of UCC that draws on the rich cultural repertoire of the Chinese working class, both past and present. When the Foxconn tragedies happened, this network of grassroots labor NGOs was immediately activated. Within a few weeks, advocates in the network produced a number of concerts, folk theater performances and most importantly hundreds of poems expressing their emotions at the time. These poems were particularly noteworthy not only because there were so many of them with intense emotions. They were also easily circulated through mobile phone (especially through SMS) and internet among workers and concerned citizens. All of these gave rise to a deluge of UCC in textual, audio and visual formats that were shared throughout Chinese cyberspace at a time when mass media reports were banned by the authorities.

Students and scholars constituted another crucial counterforce responding to the corporate irresponsibility of Foxconn and the dominance of official media. They came from various social science disciplines—sociology, social work, political science, journalism—as well as law, philosophy, and architecture. In May 2010, Ngai Pun of Hong Kong Polytechnic University, whose book *Made in China* (2005) won the C. Wright Mills Book Award in 2006, initiated a joint investigation project to examine labor issues in

Foxconn.[40] More than sixty students and scholars from twenty universities in mainland China, Taiwan and Hong Kong joined the project. I was one of them. The investigation began in July. Several teams were deployed to examine the work and life conditions of Foxconn workers in Shenzhen, Suzhou, Taiyuan, and to compare the findings across the regions.

In October, a full report by the joint investigative team was released in Beijing.[41] For this report, I made a video "Deconstructing Foxconn," which incorporated audiovisual materials from different teams and reflected group discussions at the time. The Chinese version carried by Sina.com was viewed more than 50,000 times within a week.[42] The English version was completed in November 2010 and has been viewed more than 3,000 times up to July 2010.[43] A look at the IP addresses of the viewers indicated that they were from eighty-nine countries. During the period, this video has also been screened at various events in Europe, the US and Asia, including in key Chinese cities such as Beijing, Shanghai, Hong Kong, Wuhan and Taipei. It has also been included in the Global Union Film Shorts DVD released by the International Metalworkers Federation in Geneva, Switzerland.

In a way, this video itself has become an agent of social change beyond the imagination of its producer. But it is only one of many nodes in the formation of alternative network labor beyond the control of network enterprise and the network state, which have been motivated by the exploitation of Foxconn workers. Another much more influential node in the English-language cyberspace is the Mike Daisey interview by TechCrunch TV.[44] The blowback against Foxconn is, of course, not limited to internet videos. Leading magazines like *Nanfang Zhoumo* (*Southern Weekly*) continued to report on Foxconn survivors despite official and corporate efforts to marginalize the issues. *Wired* magazine also published an article about Foxconn workers as its cover story on 28 February 2011.[45]

The most surprising development in this network blowback is the story of Tian Yu, a teenage Foxconn survivor. She jumped off her dormitory building because she did not receive her wages and she could not stand the discrimination any more. Our joint investigative team visited her when she was hospitalized, where she and her parents kindly agreed to be filmed for our video project. After they went back home to the countryside, we kept in touch through SMS and QQ. In January 2010, I visited her village in rural Hubei accompanied by a student who helped her start a blog, although this attracted little attention. The suicide attempt permanently disabled Tian Yu, who now started to make beautiful slippers in the hope that she could become financially self-reliant. A journalist in Shenzhen named Chen Yuanzhong used the Sina Microblog, the most popular Twitter-like service in China, to promote her handcrafts. Within days, the messages were retweeted by China's top microbloggers, including actress Yao Chen and reporter Luqiu Luwei. Soon Tian Yu's microblog had more than 20,000 followers and her slippers were sold out immediately (Figure 11.2).

Figure 11.2   Handmade slippers produced by Tian Yu, a Foxconn survivor, whose handcrafts became well-known through Sina Microblog, the most popular Twitter-like service in China (photographed by author).

CONCLUDING REMARKS

As the story of Tian Yu demonstrates, networks of technology and social connectivity are not always instruments for the goals of network enterprise and the network state, be they profit maximization or power consolidation. The internet and mobile phone—including the iPhone—are also tools for empowerment, alternative grassroots movements and the re-making of network labor. To see this happen, and to see ourselves as part of this crucial historical process toward a more inclusive and equitable network society, it is imperative to rethink the fundamental questions of labor in the new technosocial contexts of today.

No matter how "immaterial" the iPhone culture seems to be, its material dimension is always indispensable, depending, first of all, on the physical labor of Foxconn workers. Labor is also integral to the research and development of the iPhone, the production, testing, and installation of its software, and even the consumption of the fingertip economy. Labor is not a thing. It is a perspective. Only by using this perspective can we start to treat the workers with dignity, and extend this dignity through our activities online and offline. Only by so doing can we hope to witness the emergence of a new iPhone culture that contributes to progressive social change and a sustainable mode of network labor, beyond the shadow of Foxconn.

NOTES

1. David Barboza, "iPhone maker in China is under fire after a suicide," *New York Times*, 26 July 2009: B1.
2. Nick Webster, "Welcome to iPod City: the 'robot' workers on 15-hour days," *Mirror*, 14 June 2006: 24.
3. SACOM (Students & Scholars Against Corporate Misbehaviour), "Workers as machines, military management in Foxconn," http://sacom.hk/wp-content/uploads/2010/10/report-on-foxconn-workers-as-machines_sacom3.pdf (accessed 29 March 2011).
4. Manuel Castells, *The Rise of Network Society* (Oxford: Blackwell, 1996); Manuel Castells, *The Power of Identity* (Oxford: Blackwell, 1997); Manuel Castells, *The End of Millennium* (Oxford: Blackwell, 1998).
5. Manuel Castells, Mireia Fernandez-Ardevol, Jack L. Qiu, and Araba Sey, *Mobile Communication and Society: A Global Perspective* (Cambridge, MA: MIT Press, 2007).
6. Castells, *The Rise of Network Society*.
7. Gary G. Hamilton, *Asian Business Networks* (Berlin: Welter de Gruyter, 1996); Henry Wai-Chung Yeung, *Chinese Capitalism in a Global Era: Towards Hybrid Capitalism* (London: Routledge, 2004).
8. Jack Linchuan Qiu, "Network labor and non-elite knowledge workers in China," *Work, Organization, Labour & Globalisation* 4.2, 2010: 80–95.
9. Castells, *The End of Millennium*.
10. Castells, *The Rise of Network Society*.
11. Marcus Franda, *China and India Online: The Politics of Information Technology in the World's Largest Nations* (Lanham, MD: Rowman & Littlefield, 2002); Robyn Meredith, *The Elephant and the Dragon: The Rise of India and China and What It Means for All of Us* (New York: W.W. Norton & Co., 2007).
12. Castells, *The Rise of Network Society*; Castells, *Communication Power*.
13. David Barboza and Keith Bradsher, "In China, labor movement enabled by technology," *New York Times*, 16 June 2010: B1.
14. Maurizio Lazzarato, "Immaterial labor," in Paulo Virno and Michael Hardt, eds., *Radical Thought in Italy: A Potential Politics* (Minneapolis, MN: University of Minnesota Press, 1996), 132–146.
15. Julian Kücklich, "Precarious playbour: modders and the digital games industry," *Fibreculture* 5, 2005, http://five.fibreculturejournal.org/fcj-025-precarious-playbour-modders-and-the-digital-games-industry/ (accessed 1 May 2011).
16. Jack Linchuan Qiu, *Working-Class Network Society: Communication Technology and the Information Have-Less in Urban China* (Cambridge, MA: MIT Press, 2009).
17. Jack Linchuan Qiu, "Shanzhai culture in the network age," *Twenty-First Century* (in Chinese), 2009.12, 2009: 121–139.
18. Yu Hong, "Will Chinese ICT workers unite? New signs of change in the aftermath of the global economic crisis," *Work, Organization, Labour & Globalisation* 4.2, 2010: 60–79; Jack L. Qiu, "Network labor and non-elite knowledge workers in China."
19. Dan Schiller, "Poles of market growth? Open questions about China, information, and the world economy," *Global Media and Communication*, 1.1, 2005: 79–103.
20. Larry Hsien-Ping Lang, *Why Our Days Are So Difficult* (in Chinese) (Hong Kong: Chunghwa Books, 2011).

21. Yimin He and Rongpin Jiang, "Reform and transition: survey on labor-capital relationship in private enterprises of Guangdong Province," *China Demographics* (in Chinese), 2010.1, 2010, http://www.chinavalue.net/Article/Archive/2011/4/4/194750.html (accessed 20 April 2011).
22. Workers' Forum, "IBM Sweatshop in Dongguan: Labor Contract Law Non-Existing" (in Chinese), 7 August 2009, http://grbbs.net/thread-3296-1-1.html (accessed 16 April 2011).
23. Hong, op cit., 2010.
24. Xiangzhong Wang, Chang Cao, Yong Wang, Xianping Dong, Yiren Xia, and Dong Huang, "Migrant workers: 20 million went home due to job losses during financial crisis," *China Economic Weekly* (in Chinese), 2 March 2009, http://finance.people.com.cn/GB/8889165.html (accessed 16 April 2011).
25. Songbo Zhou and Jun Li, "Labor-capital standoff behind the bloody case in jewelry factory," *Southern Metropolitan Daily* (in Chinese), 31 March 2009: A18.
26. Lu Hua and Tingting Luo, "Please sentence me to death," *Southern Metropolitan Weekly* (in Chinese), 11 September 2009, http://www.nbweekly.com/magazine/cont.aspx?artiID=9470 (accessed 1 May 2011).
27. Austin Ramzy and Jessie Jiang, "The Chinese worker," *Time Magazine*, 16 December 2009, http://www.time.com/time/specials/packages/article/0,28804,1946375_1947252_1947256,00.html (accessed 29 April 2011).
28. Tianming Xu, *Terry Kuo and Foxconn* (in Chinese) (Beijing: China CITIC Press, 2007).
29. Interview near Foxconn's Longhua campus, Shenzhen, Guangdong Province (July 2010).
30. ATJ (Association of Taiwan Journalists), "Protesting the suppression of press freedom by Terry Kuo, Chairman of Hon Hai" (in Chinese), 30 November 2004, http://www.atj.org.tw/newscon1.asp?number=496 (accessed 15 April 2011).
31. "The stark reality of iPod's Chinese factories," *Mail on Sunday*, 18 August 2006, http://www.dailymail.co.uk/news/article-401234/The-stark-reality-iPods-Chinese-factories.html (accessed 1 May 2011).
32. Jenny Chan and Ngai Pun, "Suicide as protest for the new generation of Chinese migrant workers: Foxconn, global capital, and the state," *The Asia-Pacific Journal: Japan Focus*, 37.2, 2010, http://www.japanfocus.org/-Jenny-Chan/3408 (accessed 15 March 2011).
33. Sohu, "Who is iPhone girl?" (in Chinese), http://news.sohu.com/s2008/iphonegirl, (accessed 12 April 2011).
34. Juicun Zu, "Foxconn playing with human capital successfully: large-scale hiring after laying off 40,000 workers," *China Information World* (in Chinese), http://www.ciw.com.cn/kggs/200903/20090323143246.shtml (accessed 1 May 2011).
35. Ibid.
36. Chan and Pun, op cit.
37. Qiu, *Working-Class Network Society.*
38. Barboza and Bradsher, op cit., B1.
39. Fang Lan, "Wave of strikes in Dalian involved 73 enterprises since May" (in Chinese), *Caing.com*, http://finance.ifeng.com/news/special/cxcmzk/20100919/2636472.shtml (accessed 1 May 2011).
40. Ngai Pun, *Made in China: Women Factory Workers in a Global Workplace* (Durham, NC: Duke UP, 2005).
41. SACOM, "Workers as machines, military management in Foxconn," 12 October 2010).

42. "Deconstructing Foxconn," *Sina*, http://video.sina.com.cn/p/tech/it/v/2010-10-09/145461155789.html (accessed 19 May 2011).

43. Jack Qiu, "Deconstructing Foxconn," *Vimeo*, http://vimeo.com/17558439 (accessed 12 July 2011).

44. Andrew Kenn, "Why journalists aren't reporting the real story about Apple and Foxconn (TCTV)," *TechCrunch*, 1 February 2011, http://techcrunch.com/2011/02/01/the-real-story-apple-and-foxcon/ (accessed 1 May 2011).

45. Joel Johnson, "One million workers. 90 million iPhones. 17 suicides. Who's to blame?" *Wired*, March 2011, http://www.wired.com/magazine/2011/02/ff_joelinchina/all/1 (accessed 1 May 2011).

12 iPersonal
A Case Study of the Politics of the Personal

Larissa Hjorth

INTRODUCTION

The evolution of the mobile phone from communication device to an expressive multimedia tool has resulted in some dramatic transformations in how we experience and conceptualize the politics of labor, creativity and place. From camera phones to geomedia such as *Foursquare* (a location-based social networking software game made for the mobile phone), mobile media provide a variety of tools for the everyday user—so much so that mobile media has often been evidenced in "people power" revolutions through what Howard Rheingold calls "smart mobs."[1] While such locations as Tokyo and Seoul have long been home to the convergence of mobile and social media,[2] more recently this media marriage has become more pronounced in the all-pervasive rise of smartphones.

This birth and burgeoning of mobile media has been accompanied by the rise of the active, producing user (rather than passive consumer)—what Axel Bruns calls the "produser."[3] Mobile media, as an extension of domestic technology, has illustrated the unbounded nature of the home and domestic.[4] In a period marked by the geographic and electronic mobility afforded by mobile media, we see that one is still very much tethered—psychologically and emotionally—to a sense of home and belonging.[5] The growth of the mobile phone into mobile media is not only demonstrative of the localized and unbounded nature of the domestic, it also signals the rise of personalization. As a key attribute ensuring the success of media technologies, personalization occurs across two, sometimes-conflicting levels—user and industry. This tension takes various cartographies within mobile media. With the growth of location-based services (LBS), mobile media sees an overlaying of the social and emotional with the geographic and electronic, that, in turn, create forms of mobile intimacy that reflect what Eric Gordon and Adriana de Souza e Silva call "net locality."[6] As I will suggest, the politics of personalization are paradoxical to say the least, especially augmented through the iPhone.

Companies such as Apple, through iPhone personalization such as applications (apps), have attempted to re-brand the mobile media *evolution* as part of their "*revolution*."[7] Apple's campaign spearheads what social media expert Clay Shirky identifies as a hijacking of the personal from a realm *between people* to a *subset of technologies* in the form of personal technologies.[8] Indeed, a key role in the success of mobile media, iPhone or otherwise, has been the role of personalization. Personalization is central to the iPhone's success, largely reliant on the plethora of apps for what appears *almost* everything. But beyond the iPhone's embrace of personalization par excellence, other, more salient stories can be found to explain the personalization phenomenon—that is, the story of user-created content (UCC).

In locations like Tokyo that have been accustomed to mainstream convergent social and mobile media for over a decade through i-mode, UCC has played a key role in its success.[9] Introduced around 1999 in Tokyo, i-mode ensured the success of Japan's leapfrog into mobile internet.[10] As both a service provider (NTT DoCoMo) and a device with its own applications, i-mode provided a "personalized," walled version of the internet. Although the "normal" internet could be accessed, the device was designed so that only approved sites (following NTT regulations) could be accessed conveniently. i-Mode, like the iPhone, by way of its built-in personalization, sought to counter user personalization—a practice that has its genealogy in hacking and innovation.[11] After all, if industry could personalize the device and services in such great detail, surely users wouldn't need to alter the technology? However, such logic goes in the face of the history of mobile communication that has been synonymous with the rise of the active user and the re-purposing of applications such as SMS (short messaging service).[12] For example, SMS was never envisaged to be as big as it became; rather it was only intended to be a minor application (and specifically useful for deaf people).[13] Moreover, such revolutions as the high school pager revolution in Tokyo (in which high school girls hijacked the pager once designated for the businessman) evidenced how technologies could be adapted in ways industry never imagined.[14] This history of media subversion is particularly the case in terms of UCC—a cultural and political practice that reflects both older and emergent communication rituals.

The evolution of mobile and social media has been accompanied by, and contingent upon, the tension between industry customization (a top-down production) and user-driven practices centered around UCC and UGC (user generated content). Japan's mobile media culture, before and during the i-mode revolution, has provided many example of these two tensions playing out.[15] In this tension, there is a need to differentiate between the sometimes-interchangeable notions of UCC and UGC and their attendant forms of immaterial labor—creative, affective, emotional and social. While the former denotes the user's agency in the creation process—that is, the user becomes a "produser"[16]—the latter is marked by the user's role as a node in the circulation process. In short, UCC is made by the user, while UGC is circulated by the user. Although both involve agency, UCC is overtly creative with the *content*, whereas in UGC the user is only involved in setting the *context*. UCC and UGC are linked to notions of the personal—especially as personal technologies like social and mobile media further amplify this dimension of the user's context. Through social and mobile media—exemplified by personalization—we can see emergent types of personal politics that are inflected by the local.

With the convergence of social and mobile media—typified, if not branded by the iPhone—we have witnessed a variety of emergent and immersive personalization practices. iPhone's cornucopia of applications for almost every personal need typifies this move towards what Scott Campbell and Yong Jin Park call a personalization society.[17] But beyond iPhone's branding of this phenomenon as a subsidiary of Apple, we can see that its significance is that of a timely snapshot on the politics of personalization today. The irony is that rather than lend itself to the hacking and subversive tendencies so familiar in much of the early deployment of ICTs (information and communication technologies)[18] and social mobile media, the iPhone has been very much a closed platform device[19] akin to its often unknown precursor, the Japanese i-mode.[20] Although the iPhone is predominantly a device for accessing the web rather than also a service provider like NTT DoCoMo's i-mode, many similarities can be found—especially in terms of developing applications for customizing the web. Once an activity attributed to the personalized and often subversive techniques of the user as they adapted the technology into their everyday lives, it is the promise of the personal (made by industry rather than users) that sells many of the apps on the iPhone. So apart from an obvious re-appropriation of the older prototype of personalized mobile media in the form of Japan's i-mode, how can we begin to conceptualize some of the paradoxes surrounding the iPhone? Beyond the "newness" of the iPhone phenomenon, can we define iPhone-specific media practices? Does the iPhone produce its own affective culture?

In this chapter, I contextualize the iPhone as part of broader cartographies of personalization.[21] I want to consider the iPhone as part of the mobile media evolution and what this phenomenon—and its attendant forms of personalization—suggests about the relationship between users and media practice. Specifically, I reflect upon the iPhone as symptomatic of contemporary work/leisure paradigms by rehearsing particular dimensions associated with "personal" and "domestic" media. In order to do so, this chapter will firstly discuss the iPhone—as indicative of mobile communication and personalization practices today—within the context of approaches towards social technologies in order to sketch the methodological limits. In particular, given the increasingly geographic mobility of personal technologies, I suggest methods such as the domestication approach require revision.

Moreover, as women have historically and, even now, continue to play a key role within the domestic sphere as prime carers of the children and housework within the Australian context, it seems pertinent to focus upon women and how personalized media such as the iPhone reflect upon intimacy, work and presence. I discuss a case study of female iPhone users in Australia as they grapple with their relationship between work and home, focusing upon life after their iPhone acquisition. In this study I meditate upon the stage after the "honeymoon" phrase—that is, after users recover from the initial excitement, frenzy and novelty of the newness of apps intoxication—and how the iPhone "settles" and becomes domesticated within everyday life. This case study of female users seeks to explore some of the realities of media practice and agency as part of broader shifts within personalization.

SOCIALIZING TECHNOLOGIES: APPROACHES AND THE MOVE TOWARDS PERSONALIZATION

In the transformation from twentieth century broadcast media (like TV) to twenty-first century participatory and UCC media, notions such as the personal take on more significance in the networked environment. According to Campbell and Park, the shift toward a new "personal communication society"—epitomized by mobile communication—can be "evidenced by several key areas of social change including symbolic meaning of the technology, new forms of coordination and social networking, personalization of public spaces, and the mobile youth culture."[22] Unquestionably, through networked mobile media, how we experience intimacy, place and community is shifting. As Nancy Baym identifies, it is important to contextualize the role of new media and its so-called impact upon relationships and intimacy today, within broader media paradigms.[23] While personal technologies like the mobile phone are rewriting the relationship between mobility and intimacy, it is important to recognize that the intimate co-presence enacted by mobile technologies should be viewed as part of a lineage of technologies of propinquity.[24] Mobile media rehearses upon older social rituals as they also depart from the previous media by providing various modes of networked visual and aural communication with greater affective personalization. Through the lens of personalization, we can gain much insight into contemporary relationships and notions of place.

For David Morley, mobile technologies encapsulate some of the paradoxes of what it means to experience place today.[25] Mobile media require us to reconsider approaches such as domestication in light of the relocation of many domestic media technologies—often called personal technologies—outside the physical, bounded place of the home. However, as Morley identifies, despite this physical relocation, mobile media tether the user to notions of the home at various levels: emotional, social and psychological to name a few. Just as intimacy has migrated into the public sphere,[26] so too has the notion of the personal taken on new cartographies, reflecting the contours and textures of a "personal communication society."[27] These shifts highlight the ways in which the idea of the "domestic"—through the conflation of domestic and personal technologies and spaces—has become increasingly mobile across technological, geographic and socio-economic terrains. In turn, this requires a reworking of such approaches as domestication given that the dynamism of the home is now all-pervasive outside the physical space of the home. Moreover, given that mobile media have grown to become a main portal for the internet—and thus the main context and repository for the online and geomedia (location-based services overlaid onto social media via mobile phones)—this requires us to also consider the methodologies on offer in the burgeoning area of digital ethnography.[28] Once an approach adopted by anthropologists to study the internet like a culture, with the rise of internet and gaming studies, we see the adaptation of this approach into many subsets such as "online ethnography."[29] Given the dynamic nature of twenty-first century participatory media in which the user takes on a central position, we need robust models for approaching this phenomenon. In order to understand current shifts toward a personal communication society, a revision of some of the most formative approaches in the social technologies tradition is needed.[30]

Within the Social Construction of Technology (SCOT) perspective,[31] there have been three main approaches—substantive, affordances and social constructivism. The substantive approach was popularly known in the form of its subset—technological determinism or "media effects." This model was criticized for its simplistic understandings of technologies and users, especially negating the multi-dimensional agency of the user—in other words, ignoring the context in which the technologies/media were deployed and how this had an impact on the active ways in which the user participated and even ignored the media. Although the model was inadequate for exploring how and why users engaged in media in complex and diverse ways, it was very popular with those wanting to blame new media for social problems like violence. Marshall McLuhan was often described as working under this approach and this method gained currency again with the rise of cyberculture studies.[32]

In order to address the problems associated with the first approach, an inverse model was outlined—social constructivism. Early pioneers in the area of media technologies such as Raymond Williams influenced and informed what would become the SCOT approach. The social constructivist approach can be found in anthropology, cultural studies and sociology. Subsets of social constructivism include the domestication approach, digital ethnography (i.e., the internet as a culture[33]) and subcategories such as online ethnography (i.e., social and new media on the internet such as You-Tube studies).[34] As technologies have become increasingly more integral in social relations and a defining feature of the user's identity and lifestyle, so too have the models in which to conceptualize the dynamic relationship between the user and their technologies needed to become more complex. Technologies operate across various levels encompassing both the symbolic and material. They extend existing rituals of social interaction whilst also providing new forms of expression and media literacy.[35] The affordances approach, developed by Donald Norman, has been widely deployed in the fields of interaction design, cognitive design and human-computer interaction (HCI). Deemed an "ecological approach" by Norman, the human-centered design method tries to consider not only the actors/users physical capabilities but also their motivations, plans, values and history.[36]

Returning back to the social constructivism tradition, I will elaborate upon one of its versions, the domestication approach. The domestication approach has two strands—the British and the Norwegian.[37] The British tradition, founded by Roger Silverstone (extending upon Williams's work), grew out of media studies and an interest in consumption studies and anthropology, such as Daniel Miller's work on material cultures (the study of how objects reflect the user's identity and social context). The Norwegian version had this origin in media studies, and in later work, the social shaping discussions were further emphasized. These different approaches were deployed in the evolution of mobile media and game studies, helping to locate the media within broader cultural and technological practices.

The domestication approach focuses on technologies after acquisition, technologies in everyday life, and their *symbolic role* (borrowing from anthropology, consumption and cultural studies).[38] The approach identifies new technologies as having become embedded in everyday life and household social relations (i.e., family power relations); this results in new technologies and media not only being a site (a space or context) for making meaning, but also a place in which meaning can be gleaned. In other words, objects have their own meanings that are then put into the dynamics of cultural practice, which, in turn, redefines the meanings. This is why objects take on different meanings depending upon context. Groundbreaking case studies, such as Paul de Gay et al.'s *The Story of the Walkman: Doing Cultural Studies*,[39] although not specifically using the domestication framework, are indicative of how pivotal this approach has been for understanding the various dimensions of lifestyle technologies within contemporary life.

The need to reconceptualize the domestication approach as technologies and notions like "household" become increasingly mobile and unbounded is the focus of Thomas Berker et al.'s wonderful anthology, *Domestication of Media and Technology*.[40] In the book, the authors identify some of the limitations in the original approach—most notably, the conceptualization of the household as relatively bounded as drawn from case studies in particular industrial societies. Acknowledging the obvious changes over the past twenty years, the household is becoming less and less "clearly situated."[41]

As Maria Bakardjieva observes, the boundaries of households are more porous than the original concept of moral economy that suggested stability.[42] Sun Sun Lim identifies the role of place and culture in informing the household, an element that was underdeveloped by Silverstone et al. when they were first designing the approach.[43] Drawing from a study conducted in China, Lim highlights the cultural assumptions underscoring the domestication approach in which notions like "moral economy" and "the household" were based on Western-centric models of culture and home. As Lim observes, ICTs are important devices, both materially and symbolically, in the construction of the middle class family in China. ICTs are pivotal in building and maintaining the *guanxi*—social connections. In particular, with China's one-child policy three decades on, parents view ICTs as playing a crucial role in education. *Domestication of Media and Technology* clearly identifies some of the limitations in the original conceptualization of the domestication approach. These issues are particularly pivotal in the case of models for analyzing mobile media. As I have argued elsewhere, how we grapple with mobile media's interdisciplinary background presents many questions.[44] Should mobile media be considered in terms of domesticating new media or re-mediating the domestic?

Since both approaches have been useful in addressing the dynamic, social, creative and procedural nature of mobile media, it seems fitting to merge these two traditions by interpreting mobile media within a "domesticating new media" paradigm. However this "domesticating new media" approach fails to grasp the geosocial traversing of online terrains, now mobilized by mobile devices like iPhones and with applications like *Foursquare*. Hence, I argue that we need to deploy a hybrid approach that considers the dynamism of domestication while also engaging the fact that "the home" has migrated to new premises that are both online and co-present. One way is through a hybrid model of domestication and digital ethnographic approaches.

Digital ethnography has emerged in response to the uptake of digital media in many cultures, both developed and developing. In the rise of digital ethnography,[45] there can be found various subsets: "online ethnography,"[46] "virtual ethnography,"[47] "netnography,"[48] and "cyberethnography."[49] Although all these approaches are similar in their perception of the internet as a cultural space, they differ slightly in their specific methods or focus. One of the key ongoing factors, of which ethnographers need to be continuously reflexive, is the ethnographer's *participation*. Participation has become a central tenet in understanding mobile media personalization. In the following section, I discuss some of my findings of a case study of iPhone users in which I try to tease out some of the paradoxes today surrounding the personalization practices epitomized by the iPhone.

iPHONE PERSONALIZATION: A CASE STUDY OF USERS

In 2007 *Time* magazine named the iPhone as the invention of the year, stating that, "it's not a phone, it's a platform" that is "but the ghost of iPhones yet to come."[50] The escalation of the iPhone's popularity has been very much due to its customization in terms of applications. Paralleling the customization of Japan's i-mode in 1999, the iPhone's success has been contingent upon the applications and, in particular, the way in which Apple has managed to re-brand personalization as if it was an Apple invention. Yet the personalization of technology had been occurring long before mobile media and the iPhone phenomenon. Indeed, countries such as Japan have excelled globally in their ability to spearhead the "personal technologies" revolution from the Sony Walkman onwards. Mizuko Ito, for example, argues that it is the notion of the "personal," along with the pedestrian and portable, that has characterized Japanese technologies for decades.[51]

Part of this success is a result of their deployment of high-level personalization, particularly apparent in what anthropologist Brian McVeigh has called "techno-cute"; that is, the usage of the cute to make "warm" and "friendly" the coldness of new technologies. Personalization has been about domesticating, locating and humanizing new technologies; it is this ongoing and localized practice that is indicative of the tension between users and industry in the rise of personal technologies as intimacy increasingly becomes public.

As Goggin notes, although Apple has been slow to adopt affordances for disabled users such as the blind, it has been quick to brand a "new form of mobile consumption."[52] Like the i-mode in Japan, the iPhone plays up to a particular version of customization and personalization that highlights the user-created, industry-created tension. The evolution of the mobile phone into mobile media has been one in which customization helps the embodied experience of the media and its role as an extension of the user's identity and culture. But with iPhone applications, users no longer need to hack, as arguably all the tools they need for personalization are available as apps— supposedly. However, as Goggin observes, "The adaptation represented by the iPhone is about adapting the cell phone for the internet."[53] Here, the mobile culture of the phone is, as Goggin says, "put at the center of computing, the internet and digital culture."[54] He notes three main features: firstly, it is "a platform for, and creature of, its applications"[55]; second, it is a haptic adaptation; third, it promises to make mobile media even more "customizable and adaptable" in the practice Goggin aptly calls "identity on the move, made to order."[56] But as Goggin notes, inside this rhetoric of ultimate personalization is a paradox in terms of actual user agency in the customization process, in addition to which Apple masks the way in which the iPhone actually borrows more from previous mobile phone culture than Apple concedes.[57]

The closed platform world offered by the iPhone has very much borrowed from that offered by its precursor, the i-mode in Japan. As Harmeet Sawhney notes, the closed world of the i-mode provided Japanese users with a gated community version of the internet.[58] This paradox of personalization in the iPhone's precursor—a tension between user creativity and openness and industry's commercializing of personalization in the form of applications—is identified by Sawhney. For Jonathan Zittrain,[59] the iPhone is part of a constellation of devices he calls "tethered applications." According to Goggin's interpretation of Zittrain, "tethered devices" fail to be generative platforms and, what's more, they are configured to be actively inimical to user experimentation and co-creation.[60] As Zittrain notes,

> These tethered applications receive remote updates from the manufacturer, but they generally are not configured to allow anyone else to tinker with them. . . Indeed, recall that some recent devices, like the iPhone, are updated in ways that actively seek out and erase any user modifications.[61]

For Goggin, part of this shift lies in the way in which Apple managed to brand the smartphone evolution into an Apple revolution. As he notes,

> Perhaps what most distinguished the iPhone from many other adaptations of cell phones was its rapturous reception, and, hand-in-hand with this, Apple's phenomenally successful marketing campaign. Herein lies the paradox of adaptation that the iPhone represents. The iPhone is clearly an adaptation of the cell phone. As *Wired* magazine's Geekipedia points out, the iPhone is an obvious descendant of the smartphone—the multimedia cell phone that combines various computer programs with entertainment options. Yet the 'biggest launch since the Apollo' rebadges this evolution as a revolution.[62]

The paradoxes surrounding this re-purposing and relocation of the mobile media evolution into the iPhone revolution are perhaps best understood by moving beyond media images and rhetoric to the micronarratives of everyday users. In particular, I am interested in exploring one of the most indicative features of personal technologies today—that is, beyond the initial excitement of "newness" does the iPhone, by way of its affordances, produce iPhone-specific media practices? Is there such a thing as an iPhone affect?

In order to explore these ideas, I conducted surveys and interviews with users in Melbourne from 2009 to 2010. Given that mobile media practice is a highly gendered activity,[63] I decided to focus upon female users. It was among the female users that I witnessed the most apparent struggles between work and leisure, home and public. Moreover, as some of the respondents had children, the issue of the role of the domestic and its relationship to the domestic technologies (and hence domestication approach) was most explicit. The group of respondents were women between 18 and 50 years of age. Initially I focused on iPhone users and non-users, as well as iPhone and other smartphone (i.e., Android) users. However, it became apparent that there was something particular to the responses from iPhone users. This partly had to do with a type of iPhone affect in which users, ironically or not, were initially interested in the device because of the branding around apps. The Apps Store was often the source of discussion. For example, the use of photo apps (from *Hipstamatic*'s "analogue never looked so good" to editing suites like *TiltShift*) seemed to ignite a general interest in photography. Games were also often a source of attraction—especially for mothers wanting to distract their children whilst on the run. However, despite the initial downloading of a plethora of apps, many of the respondents noted a decline in use after the "honeymoon" was over.

In order to understand iPhone perceptions and practices *after* the honeymoon phase, I surveyed users after they had initially bought the iPhone and the types of media practice they participated in, and then again one year later, to see how these practices had been domesticated (or not) into the users' lives. This surveying occurred when the iPhone 3 was first introduced in 2009 and then again, around the time of the iPhone 4 launch in 2010. For this study, I surveyed forty women of varying ages and across different socio-economic and cultural backgrounds in Australia. One of the first things that became apparent was the fact that many respondents either had IT experience or viewed new media as both necessary for work or as a hobby. This media literacy phenomenon is not the impact of the iPhone per se but highlights how the timing of the iPhone can be viewed as part of a broader new media landscape in which women's labor and creativity feature. As I have discussed elsewhere, gender preformativity is generational with each different generation of women placing various demands and motivations of use.[64] As Leopoldina Fortunati has observed, there needs to be more studies of gender gradations, especially in terms of age and class, to understand the role of gendered mobile media practice.[65] Through the idea of mobile intimacy—that is, the overlaying of the geographic and the electronic with the emotional and socio-cultural—as something that is local as well as formed through older media practice, we can understand some of the gendered labor practices around ICTs.

In this study, we see how the phone has become the multimedia device akin to a miniature caravan that houses all personal details, much like the function of one's home as a symbol for domesticity, privacy and family. Although this caravan of sorts affords much mobility, its symbolic weight, and wait (i.e., the temporality), operates as a perpetual reminder of the various tasks and work in need of doing. This phenomenon has been called many things including the "wireless leash"[66] whereby the mobile sets us free at the same time as it creates more limits. For example, one is "free" to roam but also available. But this miniature caravan also reveals the increasing proclivity toward working at home. Just as the intimate goes public, the public—and especially work—goes private.

The tensions between the pros and cons surrounding mobile intimacy—that is, the overlaying of the physical, geographic and electronic with the social, emotional and cultural—are amplified in the case of working mothers. For working mothers, the tethering of the phone to the domestic is clear—symbolized by the always-on mothering feature that Misa Matsuda has aptly called "mum in the pocket."[67] As Lim identifies in the case of mainland Chinese parents, many see technologies like mobile media as not only important for their children's education but also a way in which to keep a perpetual eye on them.[68] Indeed the levels and layers of tethering (that often pull) afforded through the so-called mobility of mobile media has its price. Given that many women negotiate mothering along with paid work, there is a tension between mobile media decreasing yet also adding to the daily workload. Not only is the miniature caravan an embodiment of domestic labor and tethering to home, it is also about the home office. Work, like the domestic, is perpetually carried with us through the mobile phone—summed up by Melissa Gregg's notion of "work's intimacy," derived from a study in which she followed the lives of some creative workers as they struggled to differentiate between work and leisure time and spaces.[69] Far from helping creative workers, the "presence bleed" of much of online media means that "work's intimacy" has many feeling the need to be perpetually online, often scarifying other personal areas like relationships. In this blurring between work and life, characteristic of ICT's intimate publics, it is women—as often the primary carers—who bear the brunt of the wireless leash's advantages and disadvantages. While this situation was not found in this case study, many respondents did relate to a tension between work productivity and leisure in their media practices. Despite many initially downloading many "cool" apps about creativity (i.e., photo apps like *Hipstamatic*), play (standard puzzle or simple games like *Angry Birds* and haptic games like *Balloonimals*), socializing (Facebook, Twitter, *Foursquare*) and lifestyle—it was often the basic work tasks like answering and writing e-mails and surfing the internet that featured in respondents' usage. In short, their usage of the iPhone demonstrates no clear "iPhone affect."

Unsurprisingly, one of the key features that emerged in the case study was the difference in personalization practices between those who had and didn't have children. This quality played out across a variety of personalization practices—from respondents' creativity and UCC to the types of affordances and applications deployed. In this chapter, I focus upon a few of the female respondents who were working, some of whom had children and some of whom did not. Given that all respondents either worked or had immense interest in the area of new media, they were happy to discuss the iPhone in relation to other media and work practices. What became apparent in the case study were the respondents' expectations and how this reflected a particular enclave of women grappling with the "full-timeness" of both public and private elements in their life as they attempted work/life balance.

In this juggling, we see that many of the novelties of the iPhone and its plethora of applications ostensibly dissipate into a basic form of mobile internet. It became apparent that after the frenzied honeymoon of downloading numerous applications, most used their iPhone predominantly for internet searching, e-mail and maps. These features are not unique to the iPhone, rather it signals that the iPhone is little more than an i-mode for Westerners.[70]

For one full-time university lecturer and mother of two in her forties, Sophie, the iPhone was an induction to the world of Apple. Having never had an Apple computer or laptop, Sophie's purchase of the iPhone saw her quickly acquire several Apple technologies in order to be compatible. Sophie was an exception in the study, as other respondents already had Apple items like iPods or computers when they purchased the iPhone—thus, the phone was part of that media personalization continuum. This factor is not lost on Apple, who actively market its goods as inter-compatible and distinct design artifacts as much as they are technologies. A self-proclaimed high new media user, Sophie had previously used her mobile phone for more conventional purposes: SMS, MMS, photos, alarm, calendar, voice calls and games. But like many of the other respondents, the purchase of the iPhone was also an initiation into internet usage from the phone (as opposed to the computer). She saw the always-on functionality of the internet via the iPhone as a great way to sandwich work into micro-moments.

As Sophie noted, the haptic screen, QWERTY keyboard and spell check were very useful and enjoyable even though at times she felt that if she had a "pointy" finger the screen would function better. Many of the iPhone-specific apps were not really "useful" (as she considered them to be mostly games), "but they're fun, keep the kids occupied on trips in the car or shopping, etc." She continues:

> My most memorable experience is probably related to my son using the iPhone for games the first few times—e.g., when he first played *Flick Fishing*, where you flick the phone forward to cast the fishing rod, or when he played *Pocket God* and realized he could cause an earthquake by shaking the phone, he was VERY excited! There's something about the way the phone simulates gravity, movement, etc, and the way it contains "little worlds" that react to the movement of the phone itself that is very captivating (especially for kids).

For another respondent, Katherine, a mother (of one child) working in the creative technology industries and in her early forties, the iPhone was a part of a broader media ecology in which technology has been an important part of her life for a few decades. Like the first respondent, her high media and internet usage was a significant part of her life. Katherine viewed her media consumption as a product of her very busy life, which was full of multitasking. While Sophie didn't feel the need to curb her high usage, Katherine was the opposite, noting that she even moved to the country in the hope of a sea-change in which the old patterns of technology use would be broken. But part of the challenge of changing her lifestyle and the associated media patterns was the fact that she continued to work for an IT company such that her relocation meant that she was doing more work from home. So rather than decreasing the use, Katherine's practices were relocated back into the private domain. As she notes, her intense relationship with technology—thanks to her early adopter tendencies—has been something she struggled with,

> I am an extremely heavy user of which I am not proud. It's a love/hate thing. My fingers, neck and wrists hurt. I was hoping my move to the country would help curb my overuse of technology, but I seem to be more reliant than ever. I have recently been described in a review as a "technology pioneer," which makes me feel very old and dusty. I have been directly involved in technology for most of my adult life. . . I'm not sure I will ever be able to give up my addiction, but I will continue to try.

One of the interesting features of Katherine's practices was how she wove newer and older media together. As she stated, "I still read a lot of traditional books, but I am reading more and more electronically. I first started reading books on my iPhone because I was co-sleeping with my son and I wanted to continue reading in bed." Unlike Sophie, Katherine's relationship to the iPhone was part of an Apple genealogy and the role of Apple in making technologies that were more attuned to design and art-related users. For her, her lifestyle was indivisible from her technological practice.

> My tech practice and objects used to reflect this, but I now have other priorities and do not have the time or money to keep up with the latest technology. This is something that I wanted to change since having a child. I think being a Mac user vs PC user is becoming more and more obsolete. It's not so much about hardware these days, more about software. However, I am a bit of a snob and still believe that the Mac OS is geared more directly to those users who think intuitively. I am both a Mac and PC user and would always choose Mac over PC. It's just more beautiful in my eyes.

As Katherine identifies, one of the few differences between a Mac and a PC was visual (the industrial design as well as the system's architecture). One of the key features of respondents who had other Apple products prior to the iPhone was that they firstly thought of their iPhone as an extension of other, older Apple media. Rather than saying it was an extension of the MP3 player, respondents defined it as a "glorified iPod." As Katherine noted, the iPhone represented,

> Creativity and ease of use. Not bogged down in the programmatic, easier to "make" with. Thoughtful design, in terms of lifestyle choices (e.g., eco-design). Unfortunately, the "brand" of Mac and iPhone also now means exploitative labor. Similar to that of *The Simpsons*. What started off as "independent," creative and a relatively left-of-center

company has been tainted by mass market priorities. I must admit I am disappointed in the iPhone in terms of apps and some restrictions and bugs.

As an early adopter of new media with experience and expertise in the IT industry, this respondent highlighted one of the key tensions associated with the iPhone in the so-called mobile media revolution. While the rise of mobile media was accompanied and fostered by the rise of user-created content, the iPhone's highly "personalized" version of the mobile internet—echoing that of the i-mode in Japan a decade earlier—resists this history and practice.[71] Although the iPhone isn't "gated" as such, one of the ironies about Apple products like the iPod and iPhone has been Apple's resistance to a free and open system, and this is especially the case in terms of the apps. This controversy around the iPhone as a "closed" or "open" device, as highlighted by Burgess in this collection, was discussed by some of the respondents. For example, Katherine, an early adopter of technologies for more than two decades, noted that the iPhone's proprietary and semi-closed platform was a great challenge for her as someone that liked to tinker with and personalize media. With strong skills in programming, which was part of her paid work, Katherine found the iPhone system somewhat annoying. When asked if there was such thing as an iPhone person or affect, Katherine responded,

> iPhone users are pretty much like Mac users, it's all about design ease and creativity. Not so much into the programming side of technology (PC user/BlackBerry), more into the creative user side, although you can be both I suppose. Hmm, there are people who are addicted to "app chasing" and it's seen as "cool" to use an iPhone, e.g., people who use apps like "shake-it" to turn their images into Polaroids or old instant film photographs cracks me up. This type of stuff is so cool that it's lame. Is the iPhone a Polaroid? Some says it's close. I don't. These kinds of apps say a lot about nostalgia, history and the past to me- this is a good example of how the speed of technological development causes emotional upheaval and anxiety. Status anxiety too.

Here, Katherine identifies one of the key issues surrounding media personalization as the extension of identity politics. As Katherine laments, part of her relationship to media, and especially iPhone apps, is not only about a sense of history but also one's role in narrating oneself online and what this reflects about one's identity. Katherine highlights a key paradox in the agency around personalization—personalization takes much time and also shapes, at the same time as it's shaped by, one's identity. The control of re-creating one's history promised by personalization also creates much anxiety for the user. So is this (status) anxiety so prevalent to contemporary media practice amplified in the case of the iPhone or do all smartphones signpost this tension? According to Katherine,

I've never used Android, but it scares me that Google owns it, which is totally paranoid. I don't know much about it; I guess they're more driven towards gamers and people who like to develop their own apps, but I do like the concept behind it, i.e., a more open-ended platform, based on Linux. I'm relatively happy with the iPhone and don't feel like changing interfaces just yet. I love the tactile nature of the iPhone.

Here, while Katherine acknowledges that ideologically she is interested in Android smartphones by way of their open system, she identifies her relationship to the iPhone as one ordered by the haptic nature (touchability) of its interface. Indeed, the iPhone has been quick to push the possibilities of haptic media, especially in terms of games. So, has this new appreciation of the haptic through the iPhone changed Katherine's relationship to the internet and communication? Katherine notes,

I guess the biggest example would be the "separatist" nature of the "internet" on the iPhone, e.g., different apps for Facebook, Twitter, YouTube, Browser, Amazon, ABC for Kids etc. Another example is reading. I read novels on my iPhone via my Kindle app. Also, dialing into my work PC is huge, especially if stuck in a train delay. My laptop is too old and cumbersome to travel with. The battery doesn't charge for very long . . . Actually in all honesty, if I didn't have an iPhone, I think I would be making more of an effort to upgrade my laptop . . . I am constantly connected now and can use it at anytime I need to. Except my kitchen, which has no access (thank god). The biggest timewaster (fun) I have is lying on the couch and tweeting at the TV, especially *Q&A* (a political TV show on the ABC), which is GREAT FUN. It's like collective couch-potato goodness; we're all on the couch and we're all bagging Australian politics/personalities etc.

Here, the picture Katherine paints is one in which the iPhone is with her at all times and becomes an active player in all the various activities she does within and outside the home. From work e-mails to tweeting in response to a TV program, the iPhone plays a key role in her life. But when we look at her actual activities, apart from the proclivity toward the iPhone's version of the haptic, there isn't anything particularly "iPhone-ish" about her usage. Another respondent Kate, in her middle thirties and in a relationship but without children, and also involved in the IT industry, viewed it as part of her work to have both an iPhone and an Android phone to test the different phones' media and affordances. After a little while, she found herself preferring the Android phone over the iPhone due to functionality and hackability. Unlike Sophie who was a PC user and then was converted to Mac via the iPhone, Kate was a self-professed Mac "purist" who nevertheless preferred the Android over the iPhone. As Kate noted, her media practices were always "a combination of creative expression and everyday communication. In fact the two are intrinsically tied for me . . . I am a Mac user . . . purist . . . ordered . . . aesthetically pleasing . . . just how I like to organize my life." When asked about the work/life blurring and whether she constructed boundaries between the two, Kate stated,

> Being in a creative field often means that there are no boundaries between work and life. It would be like trying to stop an artist from thinking after 5 p.m. I tend to manage work "communication" by switching off after a certain time of day.

When asked about what she used the iPhone and Android phone for, as opposed to previous phones, she noted, "I watch a lot more entertainment—I either access online shows via YouTube or podcast them via iTunes. I spend a lot more [time] using it for leisure rather than just work." Her main usage was for e-mail, GPS, Facebook, YouTube and Twitter—all features not particular to the iPhone. Many of the applications she had downloaded she stopped using after a few weeks with only a few still being used if they were practical. She noted great frustration with the iPhone's closed development platform.

Zoe, a new mother aged thirty-nine and employed in the mental health sector, was also a self-proclaimed Mac lover. Her personal and professional life played out around the Mac/PC divide—at work she had to use the PC whereas at home she chose to use the Mac. For her, the iPhone allowed her to surf between these two worlds—between the "clever" and "intuitive" world of Mac and the "clumsy" PC. Like Katherine, Zoe loves the iPhone's particular version of the haptic screen and her iPhone accompanies her through a variety of work and leisure activities both in and outside the home. As someone who doesn't work in the IT or creative industries but who is passionate about technology, there was a need to create boundaries between work and the personal. As she noted,

> For me, I have had to create boundaries in my life between the personal and professional. I am now a lot more boundaried about what is work and what is personal than what I used to be. I think when I started working, I loved it and saw it as vocational, therefore relating very much to my personal life as well; however, over the years, I have realized that I want my work and personal lives to be quite separate, so I don't get swamped by work and potentially burn out.

The need to create clearly defined boundaries between work and life was something that the respondents grappled with as work and intimacy became increasingly mobile and performed both at home and away. On the one hand, the iPhone functioned as a well-designed mobile with internet, allowing them to be connected to work while at home. On the other hand, this flexibility also meant that many felt the pressure to always be "on" and in work mode while also doing domestic work. For Zoe, like other respondents, the main use of the iPhone was e-mails and internet on the run. Although she had initially starting using the calendar, she stopped after a few weeks and reverted back to the old mode of writing on paper. As she said,

> I only used the calendar to remind me of things, as I found that I missed having a paper diary (which I have since returned to using). I am very visual and the paper diary helps me to plan ahead and comprehend what I have to do in the week or month ahead, whereas I found that as I could not see this in the i-calendar, I could not really manage my week very well.

As she noted, the multimedia capacity of the iPhone was "probably wasted on me." Viewing the iPhone as fun, useful and about connectivity when asked to identify her favorite iPhone experience, she described the iPhone as just a "glorified iPod" (this was also noted by another respondent). As Zoe stated,

> Hmm, I would say discovering the joy of the iPod feature (there's something I use that I forgot about!) and downloading zillions of podcasts from radio national, which I now listen to every day as I ride to and from work and get myself an education of a different sort!

Unsurprisingly, one of the key factors separating patterns of use was who had children and who didn't. For those who had children, the role of media such as iPhones clearly helped to forge work/home fusions in what was an already highly multitasking environment. Moreover, many of these women did a lot of their work from home (so that they could effectively do at least two jobs: looking after the children and working), demonstrating the role of personal technologies in not only outsourcing domestication outside the physical home but also bringing work back into the home. The iPhone could be seen to be part of the casualization of labor whereby work becomes all pervasive and is squeezed into micro-moments between other activities in the home. As Gregg notes, with the rise of precarious and casual labor, personal technologies not only have been integral in providing models of freedom to escape the "oppressive banality" of the home but also, along with the rise in working hours, have resulted in a difficulty of maintaining a healthy work/life balance.[72] As Gregg identifies in her three-year case study, "New media technologies encourage and exacerbate an older tendency among salaried professionals to put work at the heart of daily concerns, often to the expense of other sources of intimacy and fulfillment." This sentiment is clearly reflected in the work conducted by sociologist Judy Wajcman—from her studies into feminism and technology to her work around personal technologies and the work/leisure erosions.[73]

Being a repository for, rather than a cause of, these work/home fusions, the iPhone does highlight how being "always on" comes at a cost. Just as mobile technologies set us free to roam wherever we wish, they also create new types of restrictions whereby one is always on call—what Jack Qiu calls the "wireless leash."[74] They, in turn, have an impact on the types of co-presence (being simultaneously here and there, online and offline) we participate in as the personal and intimate fuses with the public, and work bleeds into the private.

iPERSONALIZATION: CONCLUSIONS ON THE EVOLUTION RE-BRANDED AS A REVOLUTION

How does the iPhone both rehearse and break from mobile media's traditions of personalization? Does the iPhone signal the death of UCC? As we struggle to create boundaries in the face of erosions between work and leisure, public and private, personalization can often bear the brunt of this tension. In the paradoxes around personalization and its tethering to a sense of the local and vernacular for the user in tension with industry commercialization, is the iPhone the ultimate symbol of personalization? Most certainly the iPhone spearheads much of the contemporary zeitgeist, both positive and negative, of personalization. The iPhone, as part of broader global shifts toward smartphones, highlights some of the ways in which we need to reconceptualize the role between user and industry processes.

If there is a thing called "the iPhone affect," then it is definitely around its deployment and articulation of the haptic screen. Even more significantly, when speaking to respondents about their iPhones, we see how the device represents users' relationship to broader media practices in which the boundaries between "paid" and "home" work was perpetually being further blurred. As intimacy goes public, work gets further privatized and internalized.

On the one hand, the iPhone allowed working mothers to be always available to work demands, especially in terms of the constant e-mails. On the other hand, this work—in addition to domestic work—meant that many women were feeling perpetually stressed and pressured. Although this phenomenon isn't particular to the iPhone, it was the iPhone that, along with simultaneous decreasing mobile internet costs, has highlighted the increasing demands placed upon women. As a well-designed miniature symbolic media "caravan," the iPhone helps to further make malleable the work-home boundaries as work becomes increasingly casual and private and intimacy and home become increasingly mobile. The iPhone isn't a revolution, rather, like the i-mode in Japan a decade previously, it represents a Western version of a mobile with internet in which work's intimacy is all encompassing and whereby the role of personalization proffers both possibilities and limitations for the user. From its plethora of apps to its haptic screen, the iPhone offers one window onto mobile media in an age of personalization.

ACKNOWLEDGMENT

This research is part of a broader study into the relationship between online and offline communities in the Asia-Pacific funded by the Australian Research Council (DP0986998).

NOTES

1. Howard Rheingold, *Smart Mobs: The Next Social Revolution* (Cambridge, MA: Perseus Books, 2002).
2. Larissa Hjorth, *Mobile Media in the Asia-Pacific: The Art of Being Mobile* (London: Routledge, 2009).
3. Axel Bruns, "Some exploratory notes on produsers and produsage," *Snurblog*, 3 November 2005, http://snurb.info/index.php?q=node/329 (assessed 10 December 2007). For discussions of the mobile media active user, see Larissa Hjorth, "Kawaii@keitai," in Nanette Gottlieb and Mark McLelland, eds., *Japanese Cybercultures* (New York: Routledge, 2003a), 50–59; Larissa Hjorth, ""Pop" and "Ma": the landscape of Japanese commodity characters and subjectivity," in Fran Martin, Audrey Yue and Chris Berry, eds., *Mobile Cultures: New Media in Queer Asia* (Durham, NC: Duke University Press, 2003b), 158–179.
4. Thomas Berker, Maren Hartmann, Yves Punie and Katie Ward, eds. *Domestication of Media and Technology* (Maidenhead, UK: McGraw-Hill International, 2006).
5. Far from eroding a sense of home and kinship ties, mobile and social media are reinforcing the multiplicities of what constitute a sense of place. As social geographer Doreen Massey observes, a sense of place is more than a physical, geographic experience. Indeed mobile, networked technologies—that is, mobile media—not only transform how we experience place in everyday life, they also highlight that place is more than just physical, geographic notion. Thus, despite the burgeoning of numerous forms of mobility—geographic, technological, socio-economic and physical—as part of global forces, mobile media are helping to facilitate the significance of place. Doreen Massey, *For Space* (London: Sage, 2005). *See also* David Morley, *Home Territories: Media, Mobility and Identity* (London: Routledge, 2000).
6. Eric Gordon and Adriana de Souza e Silva, *Net Locality: Why Location Matters in a Networked World* (Hoboken, NJ: Wiley & Sons, 2011).
7. Gerard Goggin, *Global Mobile Media* (London: Routledge, 2011), 142.
8. Clay Shirky, "Here comes everybody," presented at *the Aspen Ideas Festival*, 30 June–8 July 2008, Aspen, Colorado, http://fora.tv/2008/07/06/Clay_Shirky_on_Social_Networks_like_Facebook_and_MySpace#chapter_01 (accessed 20 January 2009).
9. Larissa Hjorth, 2003a, op cit.
10. Nanette Gottlieb and Mark McLelland, eds., *Japanese Cybercultures* (London: Routledge, 2003).
11. See Jean Burgess's Chapter 3 in this collection.
12. See Gerard Goggin's wonderful study of invention and rise of SMS in "mobile text," *M/C Journal* 7, 2004:1, http://journal.media-culture.org.au/0401/03-goggin.html (accessed 10 June 2004).

13. Gerard Goggin, *Cell Phone Culture: Mobile Technology in Everyday Life* (London: Routledge, 2006).

14. Hjorth, "'Pop" and "Ma"; Tomoyuki Okada, "Youth culture and the shaping of Japanese mobile media: personalization and the *keitai* internet as multimedia," in Mizuko Ito, Daisuke Okabe and Misa Matsuda, eds., *Personal, Portable, Pedestrian: Mobile Phones in Japanese Life* (Cambridge, MA: MIT Press, 2005), 41–60.

15. Hjorth, "'Pop" and "Ma"; Harmeet Sawhney, "Mobile communication: new technologies and old archetypes," in Angel Lin, ed., proceedings of the *Mobile Communication and Asian Modernities I* in Hong Kong at City University of Hong Kong, June 2004.

16. Axel Bruns, 2005, op cit.

17. Scott W. Campbell and Yong Jin Park, "Social implications of mobile telephony: the rise of personal communication society," *Sociology Compass* 2(2), 2008: 371–387.

18. See Jean Burgess's Chapter 3 in this collection. See also Leslie Haddon, "The development of interactive games," in Hugh Mackay and Tim O'Sullivan, eds., *The Media Reader: Continuity and Transformation* (London: Sage, 1999), 305–327.

19. Jonathan Zittrain, *The Future of the Internet—and How to Stop It* (New Haven, CT: Yale University Press, 2008).

20. Harmeet Sawhney, op cit.

21. Hjorth, *Mobile Media in the Asia-Pacific*, op cit.

22. Campbell and Park, op cit.

23. Nancy Baym, *Personal Connections in the Digital Age* (London: Polity, 2010).

24. Larissa Hjorth, "Locating mobility: practices of co-presence and the persistence of the postal metaphor in SMS/MMS mobile phone customization in Melbourne," *Fibreculture Journal* 6, 2005, http://journal.fibreculture.org/issue6/issue6_hjorth.html ; Esther Milne, "Magic bits of Paste-board," *M/C Journal* 7(1), 2004, http://www.media-culture.org.au/ (assessed 10 June 2004).

25. David Morley, "What's 'home' got to do with it?" *European Journal of Cultural Studies* 6(4), 2003: 435–458.

26. Laurent Berlant, "Intimacy: a special issue," *Critical Inquiry* 24/2 (Winter), 1998: 281–288.

27. Campbell and Park, op cit.

28. Christine Hine, "Virtual ethnography," paper from *Internet Research and Information for Social Scientists Conference*, University of Bristol, UK, 25–27 March 1998; Christine Hine, *Virtual Ethnography* (London: Sage, 2000); Daniel Miller and Don Slater, *The Internet: An Ethnographic Approach* (London: Berg, 2001).

29. Tom Boellstorff, *Coming of Age in Second Life: An Anthropologist Explores the Virtually Human* (Princeton, NJ: Princeton University Press, 2010).

30. The study of the rise of new media technologies within popular culture can be mapped back to two key figures in the 1960s—British cultural theorist Raymond Williams and Canadian media theorist Marshall McLuhan. Both scholars were the first to study the "new medium" of the time, TV, highlighting the significance of this new media as a lens for analyzing contemporary culture. Williams founded the tradition that would be known as cultural studies through the Birmingham School of Cultural Criticism (BSCC), while McLuhan became a popular media figure by way of his skill for coining key phrases that would become part of English colloquialisms.

31. Donald MacKenzie and Judy Wajcman, eds., *The Social Shaping of Technology*, 2nd ed. (Buckingham, England: Open University Press, 1999).

32. McLuhan's substantive approach, often dubbed technological determinist, was the one that made studying new media accessible to the general masses

though car-salesman-like sayings such as "the global village" and puns like "the medium is the message/massage." Arguing that technologies extend our senses, McLuhan's work provided useful ways of understanding the role of technologies in different forms of human expression. A telephone, for example, extended the ear. However, this view of technology—the substantive—has been criticized for its inability to take into account the complexity of human agency and the fact that technologies are as much shaped by the cultures that house them as they shape cultural practices.

33. Miller and Slater, *The Internet*, op cit.; Hine, *Virtual Ethnography*, op cit.
34. Michael Wesch, "Digital ethnography," 2008, http://mediatedcultures.net/ksudigg/ (accessed 10 September 2009).
35. With this in mind, theorists such as Bruno Latour developed an approach that extended the substantive and social constructivist models—actor network theory (ANT)—whereby technologies and people are seen as "actors." Users affect the technology just as the technology has an impact on the user in a dynamic and ongoing relationship. Bruno Latour, *Science in Action* (Reading, UK: Open University Press, 1986).
36. Focusing on a relational rather than subjective or essentialist approach, the affordances approach has become a dominant consideration in the development of practical design within an increasingly HCI environment. See Donald Norman, *The Design of Everyday Things* (London: Basic Books, 1988).
37. Roger Silverstone and Eric Hirsch, eds. *Consuming Technologies: Media and Information in Domestic Spaces* (London: Routledge, 1992); Roger Silverstone and Leslie Haddon, "Design and domestication of information and communication technologies: technical change and everyday life," in Roger Silverstone and Richard Mansell, eds. *Communication by Design: The Politics of Information and Communication Technologies* (Oxford, UK: Oxford University Press, 1996), 44–74; Daniel Miller, *Material Culture and Mass Consumption* (London: Blackwell, 1987).
38. The word "domestication" comes from "taming the wild animal," and this was then applied to describing the processes involved in "domesticating ICTs" when bringing them into the home. The ongoing influence of the domestication approach attests to its importance as a tool for comprehending the socio-cultural and individualistic symbolic power of commodities, especially communication technologies, as re-enacting older rituals and cultural practices. For example, rituals around mobile communication often involve older cultural practices like gift giving. The technological function of artifacts pales in comparison to their symbolic weight and power. As Silverstone and Hirsch (1992: 20) observe, contemporary technological artifacts must be viewed as essentially material objects, capable of great symbolic significance, investment and meaning, while domestic technologies are "embedded in the structures and dynamics of contemporary consumer culture."
39. Paul de Gay, Stuart Hall, Jane Lanes, Hugh Mackay and Kevin Negus, eds. *Doing Cultural Studies: the Story of the Sony Walkman* (London: Sage, 1997).
40. Berker, Hartmann, Punie and Ward, op cit.
41. Ibid., 8–9.
42. Maria Bakardjieva, "Domestication running wild. From the moral economy of the household to the mores of a culture", in Thomas Berker, Maren Hartmann, Yves Punie and Katie Ward, eds., *Domestication of Media and Technology* (Maidenhead, UK: McGraw-Hill International, 2006), 62–78.
43. Sun Sun Lim, "From cultural to information revolution. ICT domestication by middle-class Chinese families," In Thomas Berker, Maren Hartmann, Yves Punie and Katie Ward, eds., *Domestication of Media and Technology* (Maidenhead, UK: McGraw-Hill International, 2006), 185–201.

44. Larissa Hjorth, "Domesticating new media: a discussion on locating mobile media," in Gerard Goggin and Larissa Hjorth, eds., *Mobile Technologies: from Telecommunication to Media* (London: Routledge, 2009), 143–159.
45. Wesch op cit.
46. Boellstorff, op cit.
47. Hine, *Virtual Ethnography*; Miller and Slater, op cit.
48. Roger Kozinets, "Netnography 2.0," in Russell W. Belk, ed., *Handbook of Qualitative Research Methods in Marketing* (Cheltenham, UN, and Northampton, MA: Edward Elgar Publishing, 2006), 129–142.
49. Celia Pearce (with Artemesia) *Communities of Play: Emergent Cultures in Multiplayer Games and Virtual Worlds* (Cambridge, MA: MIT Press, 2009).
50. Larry Grossman, "Invention of the year: the iPhone," *Time*, 2007, http://www.time.com/time/specials/2007/article/0,28804,1677329_1678542,00. html (accessed 1 December 2008). For critical analysis of the politics of platform (evoked by the iPhone) see: Tarleton Gillespie, "The politics of 'platforms'," *New Media & Society* 12(3), 2010: 347–364.
51. See Mizuko Ito, "Introduction: personal, portable, pedestrian," in Mizuko Ito, Daisuke Okabe, and Misa Matsuda, eds., *Personal, Portable, Pedestrian: Mobile Phones in Japanese Life* (Cambridge, MA: MIT Press, 2005), 1–16; Kenichi Fujimoto, "The third-stage paradigm: territory machine from the girls' pager revolution to mobile aesthetics," in Mizuko Ito, Daisuke Okabe, and Misa Matsuda, eds., *Personal, Portable, Pedestrian: Mobile Phones in Japanese Life* (Cambridge, MA: MIT Press, 2005), 77–102; Brian McVeigh, "How Hello Kitty commodifies the cute, cool and camp: 'consumutopia' versus 'control' in Japan," *Journal of Material Culture* 5(2), 2000: 291–312.
52. Goggin, *Global Mobile Media*, 146.
53. Goggin, *Global Mobile Media*, 140.
54. Ibid.
55. Ibid.
56. Ibid.
57. Ibid., 139.
58. Sawhney, "Mobile communication," op cit.
59. Zittrain, *The Future of the Internet*.
60. Goggin, *Global Mobile Media*, op cit., 151.
61. Zittrain, *The Future of the Internet*, op cit., 106.
62. Goggin, *Global Mobile Media*, op cit., 142.
63. Hjorth, *Mobile Media in the Asia-Pacific*, op cit.; Leopoldina Fortunati, "Gender and the mobile phone," in Gerard Goggin and Larissa Hjorth, eds., *Mobile technologies* (London/New York: Routledge, 2009), 23–34.
64. Larissa Hjorth "Mobile spectres of intimacy: a case study of women and mobile intimacy," in R. Ling and S. Campbell, eds., *The Mobile Communication Research Series: Volume II, Mobile Communication: Bringing Us Together or Tearing Us Apart?* (Edison, NJ: Transaction books), 37–60.
65. Fortunati, op cit.
66. Jack Qiu, "Wireless working-class ICTs and the Chinese informational city," Special Issue of *The Journal Of Urban Technology* on "mobile media and urban technology" 15(3), 2008: 57–77.
67. Misa Matsuda, "Mobile media and the transformation of family," in Gerard Goggin and Larissa Hjorth, eds., *Mobile Technologies* (London: Routledge, 2009), 62–72.
68. Lim, op cit.
69. Melissa Gregg, *Work's Intimacy* (London: Polity, 2011).
70. However, for Japan, the fact that it had been slow to embrace the internet via computer meant that it could leapfrog into using it via the phone.

71. This tension between the history of mobile media in fostering user creativity and a present represented by the iPhone's commodification of creativity in terms of applications and anti-hackability is discussed in detail by Jean Burgess in Chapter 3 in this anthology.
72. Gregg, op cit.
73. Judy Wajcman, Micheal Bittman, and Jude Brown, "Intimate connections: the impact of the mobile phone on work life boundaries," in Gerard Goggin and Larissa Hjorth, eds., *Mobile Technologies* (London and New York: Routledge, 2009), 9–22.
74. Qiu, op cit.

13 Four Ways of Listening with an iPhone

From Sound and Network Listening to Biometric Data and Geolocative Tracking

Kate Crawford

INTRODUCTION

Much commentary on the iPhone has focused on its visual aspects, from the way the phone looks to how it influences the act of looking, via camera, photographic apps and various forms of augmented reality software.[1] But the iPhone is also a complex technology of listening. From its inception, it was touted as a perfect combination of a music device, a telecommunications hub and an internet communicator. These can all be understood as different kinds of listening available to the user, but beyond this, there are multiple ways in which the iPhone listens back to its user. Together, the iPhone and the user form a "listening station": where an array of activities and processes can occur that are impossible alone, depending on the nature of the accord that is reached between the human agent and the iPhone.

This is an account of four "vectors of listening" that intersect through the iPhone and beyond: described as vectors because they move through other devices, people and forms of media, and they are evolving with speed. The first two sections focus on users listening in or through the iPhone; the other two consider the ways the iPhone listens to users. In each case, it is the assemblage of user and iPhone that allows for the development of particular listening practices. It is important to note that many of these vectors cross through other mobile technologies, and that the iPhone is just one in a rapidly growing market of smartphones, tablets and other connected media devices that share some of these characteristics. Mobile phones have a complex inheritance. They display their lineage in multiple ways, revealing a patina of previous media forms and prior models. Nonetheless, the iPhone has certain particularities, sometimes intentionally produced by Apple and sometimes invoked by its place in the history of mobile devices and of the other media genres that it draws upon.

This chapter contends that the iPhone offers a productive site where we can begin to consider how listening functions on smartphone platforms, what is prioritized and what kinds of bargains are being struck. We begin with a consideration of the iPhone as a sound device and follow with an account of the ways it enables forms of network listening. Next, there is an analysis of the ways the iPhone can listen to an individual's biometric data and finally as a device of eavesdropping, where the iPhone gathers— without the full permission of the user—geolocative data about his or her daily travels.

LISTENING TO PLACE: SOUND AND THE iPHONE

The iPhone contains a conglomeration of sound technologies. From its inbuilt microphone, speaker and white headphones to the onboard iPod and its capacity as a platform for new and emerging sound software, it offers multiple avenues for listening to audio. The collection of affordances it represents is the result of many histories: technological, economic, institutional and cultural. Certain activities are naturalized through its interfaces, while other forms of listening fall to the margin. Here, we consider one element of this complex set: the role of the iPhone in placing the listener in particular forms of space, be it social, environmental or corporate-owned space.

The convergence of multiple forms of aural listening in the iPhone mimics earlier telephonic technologies as well as amplifying and augmenting them into new forms of attentiveness and perception. For the first-generation iPhone, its immediate predecessor was the iPod, a device that itself possesses considerable cultural significance and recognition. Like the iPod, the iPhone represented a portable media player, one that connected to the owner's library of stored music through Apple's walled garden: the iTunes software. The iPhone encompasses the iPod, taking on its functions as an MP3 player, while also extending its reach in many new directions. Apple's careful cultural positioning of the iPod also influenced the design and cultural reception of the iPhone. Its position was consciously shifted away from the prior associations of mobiles as business tools,[2] which often saw mobile devices kept within the aesthetic strictures generally accorded to office furniture: function over beauty. Instead, the iPhone appeared as something other, an expansive glass screen that asked to be touched and stroked, illuminated by colorful icons and marked by just a single button. The design of the device has invoked considerable scholarly attention but the emblematic, white headphone/hands-free cables have their own significance.[3]

The iPhone headphones are, on first impression, the same familiar white cords that were a distinctive part of the iPod design from its first release. They look almost identical, allowing the iPhone to disguise itself, blending in with the multiple generations of iPod in public spaces. When watching someone at a train station, headphones on and gaze unfocused, it is difficult to tell what kind of device has his or her attention. The Apple branding remains distinctive, but the specific nature of the device is unclear. The woman at the station could be listening to music or a podcast or a person speaking to her on the other end of the line; only if she begins speaking in reply will the nature of the object be suggested. There is another signal that betrays its identity: a small, almost imperceptible microphone, also white, built into the cable to collect sound.

To those looking on, the white cables invoke all the images of iPod use that had been relentlessly underscored by Apple marketing campaigns.[4] Giant iPod advertising billboards in cities such as New York, London, Sydney and New Delhi offered up colorful panels featuring a silhouette of a person dancing to music, white headphones in high contrast. The replication of their form in the iPhone invoked the urban, music and youth associations of the iPod, offering a type of tightly marketed cool that was in stark opposition to alternatives, such as the all-business gunmetal or black protrusion of a Bluetooth headset. Thus, the iPhone inherited the functions of the iPod, as well as a predominant external trace. Like the iPod, the iPhone became a site of listening, but across multiple formats: music, podcasts, voice calls, as well as a range of apps and games that record, produce and mutate sound.

A recurring theme in popular and academic criticism of the iPod was that the headphones isolated listeners from the world around them. Joseph Pitt exemplifies this position when he argues that iPod users are "antisocial beings, those who avoid human interactions."[5] In his view, "The spontaneity of the social has disappeared and the silence of the anthropoid now rules." At the most simplistic level, turning the volume up on an iPod or iPhone can insulate the user from environmental sound. But this assumes a binary relationship between being "social" in public, perfectly attentive, and being silent, withdrawn and antisocial. This belies the many gradations of inattention, of not being present, that exist regardless of the presence of MP3 players. Or, in Erving Goffman's words, it is very common that "we might not be listening when indeed we have a ratified place in the talk, and this in spite of the normative expectations of the speaker."[6] But more significantly, iPhones, and iPods before them, are participants in a more complex structuring of place, of what constitutes presence and absence, while dynamically redrawing boundaries around who and what is included and excluded. The very meaning of what it means to listen to one's location is itself in flux. As Eric Gordon writes:

> Even if one doesn't carry around an iPhone or BlackBerry, as normative understandings of situations shift to accommodate new practices, network locality operates outside of the tools that enabled the practices in the first place. The tools are themselves just a medium to address much wider cultural changes around what it means to occupy space, to be with others and to be local.[7]

In this sense, iPhones do not cut people off from their location but play a role in reconceptualizing what constitutes "the local" and how we listen to the space around us: local space, personal audio space and network space. This is not purely metaphor. There are also very material applications of the re-spatialization capacities of the iPhone, including software that turns the iPhone and its owner into a form of highly attentive listener while also altering the way local sounds and social arrangements are functioning. Two iPhone apps offer examples of this kind of meta-listening: *RjDj* and *Shazam*.

RjDj is a reactive audio app that uses "the iPhone's internal microphone to 'listen' to the noises and voices heard in your proximity to dynamically create music."[8] *RjDj* encourages listeners to use the app while walking around, hearing the sounds of the city or countryside refracted through the filters and effects of the application. It creates a compelling sensation of displacement in the real, as the everyday sounds of the environment are heard through headphones, still present but strangely modified. One is listening, but the listening experience is altered and heightened. It reverses the assumption of headphones being worn as a sign of disengagement from the immediate aural surroundings and produces new forms of immersion.

Don Ihde, writing in *Listening and Voice: The Phenomenology of Sound*, describes a scene of listening to a Vivaldi concert on record. He accounts for the way the hearing self closely produces the sound in a doubled form:

> There is, in auditory imagination, the possibility of synthesis of imagined and perceived sound . . . in this case the auditory "hallucination" is not a matter of hearing one thing as something else but a matter of a doubled sound, a synthesized harmonic echo.[9]

This evocative account functions as an eerily prescient description of the synthesized harmonic echoes of *RjDj* and also suggests the appeal of the meta-listening offered by the app. It brings forth what is otherwise an act of "auditory imagination": the re-hearing and re-spatializing of sound, producing ghostly doubles. The in-ear style of the iPhone headphones ushers the immediate environment into the auditory canal with unusual closeness and intimacy while synthesizing perceived sounds with imagined, digitally processed sound. The *RjDj* user walks through space, listening to the "hearing of space."

Finally, *RjDj* also offers the option to record "scenes": to capture the moment of listening as it is heard through the algorithmic filters of the app. These scenes can then be uploaded for others to hear, giving them access to an individual's sound experience: walking through a park in summer, drinking a beer in a pub and overhearing the surrounding conversations, or cycling through the city. This, too, offers a tantalizing suggestion of an *a priori* impossible ideal of listening: to hear another's listening. Or, as Peter Szendy and Jean-Luc Nancy ask:

Can one make a listening listened to? Can I transmit my listening, unique as it is? That seems so improbable, and yet so desirable, so necessary too.[10]

RjDj's recorded scenes cannot render the particularities of someone's individual head shape, Eustachian tubes, sound receptors, neurons, memories and the myriad contributors to the lived experience of sound. But it is, even temporarily, a sense of listening to a place as heard by another—a transmitted listening to location. This highly personal sensation of iPhone listening, almost an erotics of sound, reminds us both of the intimacies and the impossibilities of listening, and the boundaries that exist in spaces and between bodies.

Shazam offers a different kind of engagement with sound and space. As a music identification app, *Shazam* allows users to "tag" a piece of music they don't recognize. It records a short sample, then analyzes it against a database. If it finds a match, it then returns the name of the song and the artist. *Shazam* works better with particular genres and artists. One reviewer explains:

> Shazam loves current Top 40 hits, most classic rock, and indie favorites. Shazam doesn't particularly care for movie scores, obscure indie rock, surf music, or '90s vintage hardcore, and is often confused by electronica—among other things.[11]

Thus, *Shazam* assists the user, acting as an external "ear" that both listens and recognizes, but only for particular kinds of music. The classic use case for *Shazam*, or other identification apps such as *SoundHound*, is in a cafe or a bar where music is playing and an unrecognized track comes on. Often this is a social process: friends will ask who a track is by, and if no one can name the artist, then *Shazam* becomes the final arbiter. In addition to the close attention being paid to the music in the local surrounds—by the user and the app, listening and identifying—another kind of space is opening up. If *Shazam* successfully recognizes an artist, it encourages the purchase of the song from the iTunes Music Store. The listeners are directed toward Apple's corporate-owned space, which may be a vast database but still has weaknesses in regard to representing non-mainstream genres and musicians. There is a particular kind of exchange occurring here, and in return for being a knowing ear, *Shazam* is also shifting a public space of listening (such as a bar or cafe) into an iTunes-branded shopping space. This is neither an unusual occurrence, nor necessarily problematic, but it has the effect of prioritizing certain forms of music as hearable, locatable and thus purchasable, and others as unknown, ephemeral and confined to a specific moment and space.

As an audio platform, the iPhone represents the convergence of multiple possibilities for attentiveness to sound and modifying sound environments.

It works in concert with the user, offering abilities and forms of knowledge, and in doing so, the iPhone user reconfigures the act of listening in space. Some of those configurations are shaped in advance, encouraging certain pathways (economic, technological) while minimizing others. The iPhone has already become a player in a wider remaking of place, significantly altering how we understand listening, both socially and phenomenologically.

NETWORK LISTENING: SOCIAL MEDIA AND BEYOND

The iPhone, along with other smartphones, opened the mobile from being a focused telecommunications device to being a media portal: connecting to several social networks such as Facebook, Twitter and Google+. Friends, strangers, colleagues, news services, celebrities: all can be heard via a range of social platforms. Conversations, posting messages, images or videos can persist while on the move, away from a computer. Another pattern emerged: "tuning in," the habit of checking the changing feed of posts multiple times during the day. For regular users of social media via iPhones, this becomes a "discipline of listening."[12]

Back in 1912, Freud developed the concept of "evenly suspended attention," which was a technique he suggested for analysts who risked exhaustion from listening to patients for many hours per day. Instead of focusing on a single line of thought, the aim was to give equal notice to all things without selection. In a consideration of Freud's recommended state of receptivity, Jonathan Crary contends that this approach brought forth more than just a method to deal with vocal streams of information that have no clear coherence. As he writes:

> It presumes an ideal state in which one could redistribute one's attention so that nothing would be shut out, so that everything would be in a low-level focus. . . My interest here is not in any specific psychoanalytic implications, but rather in the larger cultural significance of a technique designed to impose a measure of cognitive control on an unassimilable excess of information.[13]

For what began as a technique for listening to patients' voices could later be recognized in the twenty-first century metaphorical forms of listening to networks. Jonathan Crary notes the way in which Freud presages our own era, with the current emphasis on the "compulsory consumption of 'information'."[14] The extraordinary spread of Facebook, for example, contributed to the sense of compulsory participation: for a while, to not be on Facebook was to take an unusual stand and to be absent from a dominant space of interaction. The first release of the iPhone, with its simple, single-stroke access to Facebook, compounded this sense of being forever networked. Users would habituate themselves to tuning in frequently, in case something important was missed, or if they—by delaying response—were causing offence or, worse, gradually disappearing from the minds of others. Thus came the emergence of regular "listening in," checking the activity of the feeds.[15]

This emerging habit of listening to networks throughout a day comes from already established patterns of phone use: it requires people to already be habituated to checking for text messages and calls, to already be carrying their mobile with them every day.[16] As Catherine A. Middleton writes, "Mobile device usage begets more mobile device usage, addictive or not."[17] Thus, the iPhone represented a key moment of metastasis, when an already intimate, popularized technology expanded to encompass a host of media forms, with easy access to multiple spaces of listening.

I have previously outlined different forms of listening to networks: "background listening," "delegated listening," and "reciprocal listening."[18] Background listening occurs in a social media context when commentary and conversations continue as a backdrop throughout the day, with only a few moments requiring focused attention and response. Delegated listening is evidenced when there is an outsourcing of the act of network listening to other parties: when media officers update a politician's Facebook profile, or celebrities pay an agency to run their Twitter account. Finally, reciprocal listening is when two parties both "listen" to each other in social media spaces—noting and responding to each other's comments. The iPhone is a significant agent in the ability to engage in background and reciprocal listening throughout the day, untethering the modes of social media listening from desktop environments and allowing for potentially ongoing attentiveness, regardless of location or context. In a 2011 study by Pew Internet, approximately fifty-four percent of US-based adult Twitter users accessed the service from their mobile phones.[19] In the Young, Mobile, Networked study, our survey of 1034 Australians discovered that sixty-six percent of eighteen- to thirty-year-olds access social networking sites via their phones (see endnote 15). Significantly, twenty-nine percent of those accessing social networking sites from their mobiles spent more than thirty minutes each day on those sites. These figures reflect the multiple "checking in" moments that mark out frequent social networking users: not necessarily long and sustained periods of use, or even necessarily posting content, but repeatedly and briefly listening in for the latest updates.

Another reason that the iPhone has become closely associated with a wider cultural emergence of network listening is that the device emerged alongside the mass popularization of Facebook and Twitter. Twitter was first launched in July 2006, Facebook was opened to all users in September 2006, and the first iPhone went to market in January 2007. This seven-month period represents a highly significant moment for the internet, and the kind of usage patterns that would emerge. The iPhone operating system offered one-touch access to the networks that were capturing large user bases in a very short, critical space of time.

Apple very deliberately emphasized the iPhone's role as a powerful platform for accessing the internet. When launching the original iPhone, Steve Jobs described the device as a combination of three technologies: an iPod with touch control, a revolutionary mobile phone and a "breakthrough internet communicator."[20] These three features quickly became naturalized on a wide range of smartphones. But at the time of the iPhone's release, it was portrayed as uniquely combining these elements (although many smartphones in Japan already served similar functions). Its appeal as a large-screen and full-color platform to access the internet, and social networking sites in particular, was part of the wider perception that it was "the first widespread pocket desktop computer."[21] While we can consider the iPhone as an early site of mobile network listening, it is by no means the only one. In addition to RIM's BlackBerry and Windows mobiles, Google launched the Android system in November 2007, and by 2011 it had become one of the leading smartphone operating systems globally.[22]

If "evenly suspended attention" became a necessary state to sustain therapeutic efficiency, then mobile network listening became a necessary practice to sustain connection to a range of institutions, including news and information organizations, work and family. Network listening via mobiles is significant not only for the way in which it gave people what Carey describes as a sense of "cognitive control" over constant flows of information, but also the way in which it became a normative practice: an expectation that one would be contactable, and never far from the networks. The feeling of exerting control over these data flows may well be illusory, and the ushering of work into more non-work times and spaces is an established problem.[23] But there is also an increasingly sophisticated process of managing workplace, family and friend relationships within a set of separate but often overlapping networks, as well as the development of a dispersed, low-level focus necessary to maintain a presence across multiple platforms. Mobile network listening offers us a useful approach to understanding these shifts in attention and presence.

BIOMETRIC LISTENING: THE BODY AND THE iPHONE

From being a technology originally oriented toward communication and listening to the outside world, the iPhone also functions as a device that focuses its users' attention back on themselves. Ever more intricate forms of self-listening and self-management have emerged to support an increasing reflexivity in the relationship between users and their mobile phones. This can be understood as "biometric listening."

The iPhone's capacity to support processes of self-management underlies a proliferation of productivity tools available via Apple's App Store. Apps such as *Momento*, *Daily Tracker*, *TraxItAll* and *Snaptic* provide tools for users to record their movements and activities using text, images and sound recordings as they go about their day. *ReQall* presents itself as a sort of multimedia dictaphone, a way of recording ideas before they disappear from memory. Apps such as *Remember The Milk* and *Limits* provide ways to manage to-do lists and track progress against defined goals.

Self-monitoring via the iPhone also extends to matters of personal and mental health. *DietPicture* and *MealSnap* ask users to photograph their food using the iPhone's built-in camera, and in return they are told the estimated number of calories contained in their meal. *Log for Life* and *HealthEngage* attempt to track glucose intake for diabetics, while *Asthmapolis* tracks asthma inhaler usage and maps it in geographic space, potentially offering insight into physical triggers for attacks. *Mood 24/7* and *GottaFeeling* keep track of a user's mental states over time; similarly, *Track Your Happiness* correlates mood data with other events recorded by the iPhone in an effort to work out what makes its users happy.

Fitness obsessives can use the iPhone to access the kinds of performance analysis and highly detailed data previously reserved for sports researchers and elite athletes, with tools such as *iMapMyRUN*, *iMapMyRIDE*, *RunKeeper*, *DailyMile*, *RunMonster* and *PedalBrain* offering maps of workouts, logs of physical exertion and analysis of changes in speed, distance and cadence by drawing on geolocative data and the inbuilt accelerometer.

The iPhone can also listen even when a user is asleep. One popular set of apps uses the iPhone's built-in microphone and accelerometer to monitor the quality of its owner's sleep. By placing the phone on a bed beside a sleeping subject, apps such as *Sleep Cycle*, *Sleep Phase* and *WakeMate* can record and identify the sounds and body movements associated with different sleep cycles. In the morning, the user can view a report quantifying and categorizing the various phases of sleep they have just experienced. These apps can further be configured as alarm clocks that listen for changes in sleep, so that they only activate the alarm during light sleep cycles.

Applications such as these provide an increasingly intimate technological foundation for self-analysis and self-management, offering quantitative measures of performance and improvement based on the kind of continual close-range monitoring that only an always-on personal device such as the iPhone can provide. The iPhone provides its users with a range of potentially useful analytical services, but it does so by insisting that users reconstitute themselves as mobile data collection points, with the iPhone listening for changes and updating on-board and online databases with information about the personal, social and biophysical environments they move through. In order to gain maximum value from these services, users must interact with their phone in a way that generates well-structured data. Users strike a balance between the value that they can extract from the device and the lifestyle changes the device imposes as it gathers information about them. In effect, this functions as a pact: the more structured data a person allows the iPhone to gather, the more the device can offer by way of a meaningful analysis of that person. An individual learns to speak in languages that can be heard by the iPhone; the iPhone reciprocates by listening and then responding, with richly detailed and personal stories of the user's daily data patterns and how they change over time.

Amateur athletes, who may previously have compared their performance against that of their peers and competitors, can now track their improvements against objective criteria using automated tools on the iPhone. Dieters can consult an impersonal device rather than seeking advice from people with nutritional experience. Those suffering from depression can supplement or replace the observations of their friends and family with an app that helps them keep track of changes in mood. By externalizing subjective data, users are offered the prospect of an objective understanding of various dimensions of their personal lives. Yet individuals can only realize these benefits when they subject themselves to scrutiny by the device and its on-board software. Individual activities must be measured, quantified and stored as coherent data sets. Activities must be reduced to a form that can be described in terms of discrete performance values and determinate states. When a user accesses the application's reporting functions, he or she implicitly acknowledges the validity of the results and acts accordingly, adapting to the new information by making lifestyle changes or adjusting personal goals. As such, the particular forms of rationality codified in the phone's software gradually leak out and produce effects in the lives of users.

Ultimately, while the biometric apps produce particular kinds of useful information to users, their operations are embedded within a more ambiguous legacy. Max Weber suggested that modern capitalism is characterized by the continual extension of scientific rationality into areas previously considered unknowable or uncertain.[24] Rational perspectives could now be applied to social and personal spheres previously understood as being governed by shifting subjective forces, which Weber viewed with some ambivalence. On the one hand, it offered the prospect of significant improvements in analytical capacity over new realms of understanding. On the other, it suggested that many mysterious human capacities could be irreparably damaged through their subjection to rational and bureaucratic rules, a process he designated as "disenchantment." Disenchantment in Weber's terms can be defined as:

> The historical process by which the natural world and all areas of human experience become experienced and understood as less mysterious; defined, at least in principle, as knowable, predictable and manipulable by humans; conquered by and incorporated into the interpretive schema of science and rational government.[25]

A similar observation appears in Foucault's work on the governance of the self, where he suggests that scientific advances function as instruments of power as much as extensions of knowledge.[26] In his later work, he developed a notion of "care of the self," according to which individuals make use of particular technologies to constitute themselves as individual subjects.[27] The iPhone could be understood within a broader technological system of self-management, providing mechanisms for users to reflexively develop and assess their capacities but simultaneously placing limits on the available forms of self-understanding. Of course, individuals are still able to sustain a critical engagement with the mobile technologies they use, modifying both the applications themselves and their use of them. In the Young, Mobile, Networked study, participants offered nuanced accounts of what applications they use, how and why: regarding them as an open set of options that could be regularly augmented or deleted at will. As such, users were able to modify the set of rationalities exposed by the device to suit themselves. Nonetheless, the gathering of data to suit biometric tracking apps produces another set of problems: how secure is the data, and who has access to it?

LISTENING AS EAVESDROPPING: THE iPHONE AND LOCATION TRACKING

Imagine a map of the city where you live. Tracing across it is an array of circles, in colors shifting from deep orange and red to dark blue and purple. The circles overlap and vary in size, but they form clear clusters and lines, focusing on the areas where you spend the most time and the paths you tread most regularly: perhaps from school to home, or home to work or between your friends' houses. It has an eerie appearance; you can see how ingrained your patterns of travel are and how many areas of your town you never enter or explore. It knows everywhere you've been for the last year: an externalized memory trace of thousands of small journeys.

In 2011, two researchers working for technology company O'Reilly Media, Alasdair Allan and Pete Warden, wrote a piece of software called *iPhoneTracker*.[28] This software caused major international controversy, and a public relations disaster for Apple. In essence, *iPhoneTracker* is quite simple: it extracts location data from iOS 4 (the iPhone 4 operating system) and graphs the data on a map. The result is a map that features a range of circles, which appear to trace the whereabouts of the phone user. The release of *iPhoneTracker*, however, caused alarm to spread among iPhone users, particularly once it was revealed that the latitude and longitude data of the phone's location was being backed up via iTunes to the user's computer, where it was stored unencrypted. Thus, anyone with some knowledge and access to the user's computer could view the location and times of their comings and goings. It was the stuff of good scare headlines, and duly a privacy scandal erupted.

In fact, this kind of tracking data has always been available—but only to a select few. Mobile telecommunications providers have always had access to the location data of a mobile, as it moves between mobile base stations, which can be appropriated (with a warrant) by law enforcement or obtained illegally by tapping into mobile network databases.[29] But here was a claim that Apple had installed a log file that was storing this sensitive data on every iPhone and backing it up with every synchronization, without permission of users, and in such a way that it could be readily accessed. Alex Levinson, a data forensics researcher, had previously noted the existence of the log file in iOS 3, but it was only with the advent of iOS 4 that the files began to be backed up with every synchronization with iTunes, such that the data was being recorded indefinitely.[30]

Apple responded with the claim that location data was only tracking via base stations, which could be kilometers away from the user, and that any data reaching them was anonymized. Further, Apple revealed that a bug was responsible for the infinite data collection: it should only store a week's information, as a way of improving GPS-related functions on the phone.[31] However, the company did note that it was collecting anonymous traffic data to build a crowd-sourced traffic database, aiming to give iPhone users improved traffic services. So although the scandal dissipated, the phone was still listening to users: just not as closely, nor with such lasting memory of its user's travels.

Location data is becoming increasingly valuable. Companies are seeking to own and control more location data, with the aim of on-selling niche services as well as advertising, and to gain a higher resolution picture of what their customers do and where they go.[32] But there is another kind of bargain at work here: users can make a choice whether to offer their location information (although as *iPhoneTracker* shows, there is a considerable gray area around the operating concept of "choice"). Nonetheless, there are factors at work in how these choices are made. Often users privilege convenience (and new capacities, such as live map access) over privacy.[33] But as George Danezis, Stephen Lewis and Ross Anderson have demonstrated, there are also economic decisions being weighed—users can often nominate a price for which they are prepared to give away information about their location.[34] The considerable questions that remain about the use of locative data include what kind of consent is being given for that price, let alone what kinds of uses that data will serve over time.

One of the elements of the iPhone tracking controversy that disturbed users was the permanent and unending nature of the data collection: large volumes of data, collected over years, can reveal an incredibly detailed depiction of an individual's life, associations and preferences. What might be deemed acceptable at one point in a lifetime may not be at another; information may be acceptably harvested by a company, but not if given to a government, or vice versa. The serious problems facing the rapidly expanding field of locative data—and the iPhone's role within them—is the nature of the bargains being struck, and whether all parties have full knowledge of what role they are playing and where it ends.

Finally, beyond the concerns about data use and misuse, there is an issue of aesthetics. This is where *iPhoneTracker* offers a different kind of insight. Contained within the richly colored circles, illuminating the user's path between mobile base stations of their town, was something beautiful and nostalgic: the ability to reflect on the paths and experiences of one's life, remembered with far greater reliability than most people are able to summon. Alexis Madrigal, an editor at *The Atlantic*, installed *iPhoneTracker* and visualized his data:

> Here, each little clump of cell phone pings reminds me of a story. There's the time I went to Great Falls, and another time to an Audubon bird-watching preserve, and Annapolis, and a trip down to Richmond. I can see where I travel in the city and what terrain remains unexplored.[35]

This is location data nostalgia: a genre of personal reflection that is impossible without the kind of relentless machinic listening offered by our mobile devices. Location memories have been externalized, allowing the iPhone to record for us and later remind us where we've been. The circles give a rough approximation, with the rest being left to the user's imagination and recall. Acknowledging the clear and serious privacy implications of geolocation tracking, particularly without full consent, Madrigal nonetheless pauses to draw an emotional remembrance from the data. Like Jorge Luis Borges' character Funes the Memorius, who could remember everything he saw, the iPhone is an implacable data collector. But Funes, for all his prodigious recollection, could not reflect on what he saw: "to think is to forget differences, generalize, make abstractions. In the teeming world of Funes, there were only details."[36] The iPhone, too, is tracking details of movement, but without the narratives or driving forces that animate them.

CONCLUSION

The iPhone, both culturally and historically, represents a key site to understand the development of listening practices. The four "vectors of listening" discussed above are by no means an exhaustive list, as types of listening are evolving and multiplying on smartphone platforms. But the iPhone captures a moment in time when many of these forms of listening converged in one device. From listening to music and conversations to the many apps that alter, enhance or distort our listening, the iPhone can augment the human ear as well as insulate it. As an Apple product, it is indentured to iTunes and the particular forms of commercial space controlled by the parent company. It also redraws the boundaries of what constitutes listening in public spaces. From reactive audio applications that respond to the immediate environment to music identification apps that engage with background music, the iPhone is a significant agent in the remaking of place and of the act of listening to the sounds around us.

As a platform for network listening that allows users to regularly tune in to the social media spaces where they maintain a presence, the iPhone is a site where users experience and develop disciplines of listening. Regularly checking to read ongoing updates from friends, associates, colleagues and strangers has the quality of a background channel, like half listening to a radio. Nonetheless, this kind of regular "tuning in" develops into a normative practice, where being available, present and attentive is expected from active participants in social media spaces. This is also part of the larger process of constructing particular circles of social connection: the people who are ignored, those who are listened to and those who will receive a response.

In addition to the ways users listen to and through the iPhone, the iPhone attentively listens to the user, in terms of both tracking their biometric data (through applications designed to monitor sleep, heart rate, diet and so on) and recording location data from mobile base stations whenever the user travels. One type of listening offers to assist the user—through a bargain whereby the giving up of data will return richer information over time—as the iPhone will record, analyze and remember the user's patterns. The other—eavesdropping by location tracking—is more ominous, taking information and storing it on the user's computer without clear permission. While Apple has promised to remedy the excesses of this form of listening, the concerns about location data on the iPhone iOS 4 nonetheless reveal the way these devices and the bargains we make with them can have unintended and longlasting consequences. There are already vast stores of iPhone-produced latitudes, longitudes and time stamps, stored on computers around the world. Depending on who is listening to that data, it may be used to seriously infringe privacy, to produce more targeted services and advertising or to draw out a set of personal meanings. In the broader listening process that includes both humans and iPhones, all that endless detail without context can be transformed into data with profound meaning and enduring effect.

NOTES

1. Chris Chesher, Chapter 7 in this volume; Andreas Brogger and Omar Kholeif, *Vision, Memory and Media* (Liverpool: Liverpool University Press, 2011).
2. Tomoyuki Okada, "Youth culture and the shaping of Japanese mobile media: personalization and the keitai internet as multimedia," in Mizuko Ito, Daisuke Okabe and Misa Matsuda, ed., *Personal, Portable, Pedestrian: Mobile Phones in Japanese Life* (Cambridge: MIT Press, 2005), 41–60; Catherine Middleton, "Illusions of balance and control in an always-on environment: a case study of BlackBerry users," in Gerard

Goggin, ed., *Mobile Phone Cultures* (Oxon and New York: Routledge, 2008), 28–41.

3. Katie Ellis and Mike Kent, "iTunes is pretty (useless) when you're blind: digital design is triggering disability when it could be a solution," *M/C Journal* 11(3), 2008, http://journal.media-culture.org.au/index.php/mcjournal/article/viewArticle/55 (accessed 15 June 2011); Alicia David and Glore Peyton, "The impact of design and aesthetics on usability, credibility, and learning in an online environment," *Online Journal of Distance Learning Administration* 13(4), 2010, https://www.westga.edu/~distance/ojdla/winter134/david_glore134.html (accessed 10 June 2011).

4. Kate Crawford and Gerard Goggin, "Handsome devils: mobile imaginings of youth culture," *Global Media Journal*, 2, 2008, http://www.commarts.uws.edu.au/gmjau/iss1_2008/crawford_goggin.html (accessed 12 June 2011).

5. Joseph Pitt, *Doing Philosophy of Technology* (Dordrecht: Springer, 2011), 39.

6. Erving Goffman, *Form of Talk* (Philadelphia: University of Pennsylvania Press, 1981), 131.

7. Eric Gordon, "Towards a theory of network locality," *First Monday* 13(10), 2008, http://firstmonday.org/htbin/cgiwrap/bin/ojs/index.php/fm/article/viewArticle/2157/2035 (accessed 22 May 2011).

8. Jason Kincaid, "RjDj continues to be the most trippy app on the iPhone," *TechCrunch*, 2008, http://techcrunch.com/2008/12/29/rjdj-continues-to-be-the-trippiest-app-on-the-iphone-and-i-love-it/ (accessed 20 May 2011).

9. Don Ihde, *Listening and Voice: A Phenomenology of Sound* (Athens, OH: Ohio University Press, 1976), 62.

10. Peter Szendy and Jean-Luc Nancy, *Listen: A History of Our Ears*, trans. Charlotte Mandell (New York: Fordham University Press, 2008), 5.

11. Ben Boychuk, "Shazam for iPhone," *Macworld*, 2009, http://www.pcworld.com/businesscenter/article/158158/shazam_for_iphone.html (accessed 17 June 2011).

12. Kate Crawford, "Following you: disciplines of listening in social media," *Continuum: Journal of Media & Cultural Studies* 23(4), 2009: 532–533.

13. Jonathan Crary, *Suspensions of Perception* (Cambridge, MA: MIT Press, 2001), 368.

14. Ibid., 369.

15. I draw this claim from the data generated by the Young, Mobile, Networked study, a three-year Australian research project conducted by Gerard Goggin and myself. We interviewed over 300 people around Australia and surveyed over 1000. Many talked about their varying experiences of listening to networks via mobiles. As one twenty-seven-year-old female explains why she got an iPhone, "I liked the idea of being able to indulge my compulsive urge to check everything . . . and Twitter fed my obsessive need to know things all the time."

16. This kind of habituated connection to the mobile is strongly evoked by one respondent in the Young, Mobile, Networked study: "My phone's usually not more than like three metres from me at all times . . . I even admit taking it to the bathroom with me. When I'm going to have a shower, I'll have it on the bench. It's so sad. It's under my pillow when I sleep."

17. Middleton, "Illusions of balance and control in an always-on environment: a case study of BlackBerry users," op cit., 36.

18. Kate Crawford, "These foolish things: on intimacy and insignificance in mobile media," in Gerard Goggin and Larissa Hjorth, eds., *Mobile Technologies: From Telecommunications to Media* (New York: Routledge, 2009), 252–266; Crawford, *Continuum: Journal of Media & Cultural Studies*, op cit., 525–535.

19. Pew Internet Report, 2011:http://www.pewinternet.org/Reports/2011/Twitter-Update-2011/Main-Report.aspx (accessed 10 August 2011)

20. Steve Jobs, *Macworld San Francisco 2007 Keynote Address*, 2007, http://itunes.apple.co m/us/podcast/apple-keynotes/ (accessed 25 May 2011).

21. David Pogue, "Hello Blackberry, Meet the iPhone," *New York Times*, 13 March 2008, http://www.nytimes.com/2008/03/13/technology/personaltech/13pogue-email.html (accessed 30 May 2011).

22. Tarmo Virki and Simon Carew, "Google topples Nokia from smartphones top spot," *Reuters*, 31 January 2011, http://uk.reuters.com/article/2011/01/31/oukin-uk-google-nokia-id (accessed 2 June 2011).

23. Middleton, "Illusions of balance and control in an always-on environment: a case study of BlackBerry users," op cit.

24. Max Weber, *The Protestant Ethic and the Spirit of Capitalism* (New York: Scribner, 2003).

25. Richard Jenkins, "Disenchantment, enchantment and re-enchantment: Max Weber at the millennium," *Max Weber Studies,* 1, 2000: 11–32.

26. Michel Foucault, *Madness and Civilisation: A History of Insanity in the Age of Reason* (New York: Vintage Books, 1988); Bryan Turner, "The government of the body: medical regimens and the rationalization of diet," *British Journal of Sociology* 33(2), 1982: 254–269.

27. Michel Foucault, *The Use of Pleasure* (New York: Vintage Books, 1990).

28. Alastair Allan, "Got an iPhone or 3G iPad? Apple is recording your moves," *O'Reilly Radar*, 20 April 2011, http://radar.oreilly.com/2011/04/apple-location-tracking.html (accessed 22 April 2011).

29. Although researchers at iSec Partners revealed in 2010 how to track the location of almost any GSM handset in the world, using methods that are not yet defined as illegal. See Elinor Mills, "Legal spying via the cell phone system," CNET, 21 April 2011, http://news.cnet.com/8301-27080_3-20002986-245.html (accessed 2 July 2011).

30. Alex Levinson, "3 major issues with the latest iPhone tracking discovery'," 21 April 2011, https:/ /alexlevinson.wordpress.com/2011/04/21/3-major-issues-with- the-latest-iphone-tracking-discovery/ (accessed 15 June 2011).

31. Apple, "Apple Q&A on location data," Apple web site, 27 April 2011, http ://www.apple.com/pr/library/2011/04/27Apple-Q-A-on-Location-Data.html (accessed 10 June 2011).

32. MG Siegler, "The great location land rush of 2010," *TechCrunch*, 2009, http://techcrunch.com/2009/12/23/location-2010/ (accessed 22 June 2011).

33. Alex Madrigal, "My life according to the iPhone's secret tracking log," *The Atlantic*, 2011, http://www.theatlantic.com/technology/archive/2011/04/my-life-according-to-the-iphones-secret-tracking-log/237636/ (accessed 17 May 2011).

34. George Danezis, Stephen Lewis and Ross Anderson, "How much is location privacy worth?" *Fourth Workshop on the Economics of Information Security*, 2005.

35. Madrigal, 2011, op cit.

36. Jorge Luis Borges, *Labyrinths* (Harmondsworth: Penguin, 1970), 94.

14 How a University Domesticated the iPhone

Ilpo Koskinen

INTRODUCTION

This chapter studies how the iPhone entered one formal organization, the University of Art and Design Helsinki (Taik). Like mobile phones throughout their history, the iPhone was a coveted object that people passionately wanted as soon as it entered the market. However, Taik had a mobile phone contract with a carrier that did not have the iPhone in its selection until summer 2010. This paper shows how Taik found settlements between passion and organizational policy through several routes from 2007 to 2010: at one School, through a conflict; at another, through research-based justifications; at a third, through the very reason for existence of the School.

iPHONE AS A MORAL OBJECT

One strand of literature on mobile phones has looked at mobiles as more than things for calling and texting. In what must be the first empirical study on mobile phones, the sociologist Timo Kopomaa noted that phones are moral objects.[1] Just like many other novelties, people and institutions observe and evaluate mobile phones, and people who buy them may have to explain the reasons for buying and using them.[2] Typically, buyers and users appeal to reason for buying their phones: phones are tools, not toys.[3] A few years later, Leopoldina Fortunati, James E. Katz and Raimonda Riccini edited a book that looked at phones as fashionable objects.[4] From around 2000, phones were treated as accessories, and companies like Nokia followed the fashion and luxury industries keenly to create phones that would attract a following among the fashionable set.

In 2007 and 2008, the iPhone was hotter than any other phone at that time. Some reasons were in the cult status of Apple and its American origin, but the iPhone's sleek interaction design and minimalist appearance also played a part. The iPhone, however, was not just hot. It was also a dangerous object, especially in formal organizations. People wanted it, but large formal organizations had phone policies that tied them to carriers that did not always have an iPhone in their selection. Also, the iPhone was at the more expensive end of the market, which often placed it outside the price range accepted by many organizations.

Reasons like these turned the iPhone into a possible source of conflict. Organizations work on premises that follow organizational rationalities; people work on other premises. These may work in harmony, but just as well be at odds. When they are at odds, there needs to be ways to reconcile these alternate realities, hot and cool, passion and organizational rationality. This paper focuses on smart phones in organizations for several reasons. Organizational uses of mobile phones have received little attention in mobile studies, even though they make up a major part of the smart phone market. Organizations are also good laboratory animals for studying issues like the clash between passion and rational decision-making.

MISTAKES AND RULES AT WORK

Organizations create policies that justify some actions, and make some others punishable. Members of the organization are held accountable for following these policies. Typically, everyone is expected to know these policies, but it is another question who can define which deviations are noticed and how they are interpreted.[5] In this respect, organizations are moral systems.

The American sociologist Everett C. Hughes once urged sociologists to look at mistakes at work as a way to understand power in organizations. His point was simple enough: everyone makes mistakes occasionally, but penalties are not distributed evenly. Some get away with practically anything; some have to be wary about making mistakes; some get no leash at all.[6] Following Hughes, Charles Bosk has described in his marvelous ethnography *Forgive and Remember,* how some breaches are forgiven of medical students while some others are not. Attending physicians accept technical errors as long as they are not fatal, but are wary of moral errors that reveal the student's character. Good students can get away with almost anything, while those whose character is found faulty are punished for the tiniest of mistakes.[7]

Usually however, these evaluations are based on organizational rules rather than informal character assessments. Typically, it is senior management that has the job of seeing that things go according to the rules. Their job is to monitor things and decide what kinds of moral implications they have.[8] At medical school, at stake is entry into the most prestigious medical specialties. In formal organizations, punishments vary from barely noticeable to losing one's job. Organizational rules, however, are not cast in stone. They can also be changed if senior management so decides.

In this paper, "domestication" is shorthand for describing the ways in which the iPhone enters an organization, not an explanatory concept. Literature on the domestication of technologies originated with Roger Silverstone's writings, but has focused on households, and has been largely void of questions like power and organizational rules.[9] For this reason, the start-ing point of this paper is in Hughes' and Bosk's writings.

POLICY MEETS PASSION

The University of Art and Design Helsinki (Taik) was a large design uni-versity by European standards. In 2007 to 2008, when the iPhone saw daylight, the university had a contract with Elisa, the mobile carrier of the former Helsinki Telephone Company. Everyone at the university who was entitled to a company phone was to get a mobile policy from Elisa.

When the iPhone entered the Finnish market in 2008, Apple did a pack-age deal with TeliaSonera, a company formed a few years earlier in a merger of former Telia, Telecom Sweden, and Sonera, Telecom Finland.[10] For pri-vate use, anyone was able to buy an iPhone. Charging calls, text messages and data transfers from the university was out of the question.

The management did not see any reason to change the rules: the iPhone was barely seen as a novelty in a country that had been one of the world leaders in mobile telephony since the mid-nineties. Although the iPhone had the cool factor, it was technically inferior to Symbian-based smart phones. Business solutions like maps, mail and calendar syncs were easier and safer to use in Symbians than in iPhones. Multimedia capabilities like cameras and MMS had been available from around 2002 in most phones sold in the Finnish market. In a culture like this, the management found few reasons for opening the doors to the iPhone and breaching its contract with Elisa.

Of course, there were ways around the TeliaSonera package deal. A few days after the iPhone's release, it was possible to hack the iPhone's SIM cards so that the device became carrier independent. Anyone with some technical skills and access to the Web could buy an iPhone from the United States in 2007 and hack it into a working phone. An easier way was to buy a Sonera policy and send the bill to the university.

The university, however, could refuse to pay it with a reason good enough to be defended in court. Managers in charge were tied to Elisa policy; if they departed from it, they might have been held liable for breaking the law. University policy, in effect, made it impossible to buy and use iPhones, although Taik is a design school, which mostly lives in "Apple culture." The policy and these workarounds created a situation in which many types of reasons drove people to buy iPhones, most of these reasons being based on passion and desire to have the latest Apple product, while some were legiti-mately based on professional needs—like in the case of media designers.

This was a temporary situation. It began on 11 July 2008, when the iPhone 3G came to the Finnish market, and ended in Summer 2010, when Apple's package deal with Sonera ended. Taik's mobile partner Elisa begun to sell the iPhone and iPhone subscriptions, making it an ordinary object in the eyes of the university. By then, Taik had also lost its independence and become a part of Aalto University, which also had engineering and business Schools. Aalto's phone policy made it possible from the very beginning to buy an iPhone with an Elisa connection.

The tumultuous period, then, lasted for about two years. During this period, several settlements took place at Taik. The conflict between passion and policy was solved in several ways, involving several types of processes, negotiations, and power positions. This paper looks at the ways in which three Schools of the university reached settlements.

THE iPHONE AS A MORALLY CONTESTED OBJECT

At one of the university's schools, passion for the iPhone came into conflict with Taik policy in 2008. The settlement was painful and caused serious conflict between senior administration and academic leaders. It shaped policy in this school for two years.

Two employees bought iPhones with Sonera cards and started to use the phone as a work phone. A third employee bought the phone from abroad during a business trip, hacked it and started to use it as a Taik phone. Administration at the school took notice and refused to pay for the phones when these employees handed in the receipts. The same happened when the first phone bills arrived. For senior administration, the situation was clear: a breach of rules had happened. Its position was that these three people had to return the phones because they were not acquired according to Taik policy. The school could not pay for calls and data in the Sonera network. The reasoning was straightforward: this would have been against Taik's contract with Elisa. Employees had made a mistake; the senior administration could not see this mistake through its fingers.

The employees, however, did not accept this line but appealed to the dean of the school, who then listened to both parties. From the employees, he heard a host of reasons for why the iPhone was a necessity. One of the employees, a former vice dean, justified his reasons by claiming that he needed an iPhone for teaching media production for small screens. Another employee's reason was that he needed an iPhone to maintain students' computer systems from his home. The third employee's reason was that he needed to test its sound capabilities, again with media production in mind. To the dean, then new in his job, these reasons were convincing enough to overrun the administration's line. For him, these reasons were professional enough to justify exceptions to Taik policy.

Senior administration was not happy and saw this development as evidence of moral failure: it was not anymore a mere mistake. Fearing what would happen to its authority (and the school's budget) if the iPhone floodgates were opened, it took the emerging conflict to central administration. Unsurprisingly, it supported the administration's line, and said in no uncertain terms that the phones were against company policy. The message was that they were not allowed, and they had to be returned. This response, however, posed a threat to the dean's face. He was forced to choose his side.

In terms of power, the situation was easy enough: deans have enough power and resources to run their Schools as fiefdoms. All they need to break company policy is to find a proper justification for breaking away from it.

However, this requires some balancing; a cautious dean does not humiliate his key administrators. After about six weeks, there was a simple settlement. Two employees agreed to pay their own iPhone bills and kept the ownership of their phones—that is, they did not try to charge the price of the phone to the university. One employee agreed to separate work use from personal use, and use a separate affix for his work calls to indicate which use was work-related. The school paid this part of his phone and data bill.

This was against Taik's phone policy, but the administration backed up and accepted this change. Others at the department agreed not to buy iPhones. These three phones, then, did not open the floodgates; the school policy remained restrictive. In reaching this settlement, the administration was successful in its effort to enforce the company policy, and none of the employees lost the visible token of their passion, the iPhone.

RESPECTING ORGANIZATIONAL MORALITIES

The contrast between the previous case and the School of Design illustrates the importance of the management's definition of what is normal and its power in defining the borderline of acceptable and non-acceptable errors. At this school, Taik policy was overrun in a few cases, but with academic and senior management's consent. For this reason, iPhones at this school did not raise questions of mistakes or moral failures.

Few academics at the School of Design have iPhones even in 2011, because they generally find them technically inferior to Symbians (and possibly also Androids). Deans and department heads do not have one; and only two professors—one personal, another a company phone. The attitude was restrained from the beginning. The dean did not authorize purchases without very good reasons, and no one challenged her authority.

This was the main picture, but it cracked on a few occasions. The school respected Taik's contract with Elisa but allowed a few exceptions for solid organizational reasons. The first iPhone was bought for a research project that developed interactive maps of a national park close to Helsinki. In the project, researchers from the Geodetic Institute and the university wanted to create interaction designs for the park, but Symbian proved to be too difficult to function as a proper test platform. As standard interaction techniques were far easier to implement with the iPhone, it was chosen as a technology platform.

For this reason, the then vice dean of the school authorized iPhone purchases with Sonera SIM cards for researchers in this project. In the end, only one researcher, a computer scientist, bought it. The school paid for his iPhone from project funds. In fact, in October 2008, the vice dean went to mobile phone shops with the researcher to study the alternatives and to compare the prices. They even learned about available hacks in case they would be needed.

However, the researcher bought a Sonera SIM card for his phone and did not try to charge his calls and data to the university, even though the vice dean had told him that he would approve it. The researcher said that as Sonera subscriptions for unlimited 3G data were only 12 Euros per month, and that this covered using the iPhone as a modem, it did not make sense to separate a sum this small into work and personal use. Even today, as he uses mail for communication and makes only one or two work calls a month, separating the bill would make no sense. When traveling, his habit is to seek the cheapest alternative for communications. For instance, when his present research project takes him to Cambridge, UK, he uses cable connection for his laptop and connects the iPhone to the laptop for Skype calls.

This became the bottom line within the School of Design. Breaking this policy was fine as long as the dean and the vice dean were able to find a solid reason that justified a departure from the policy. They needed a rationale to explain their decision in case someone would ask why the School did not follow university policy. This was a very restrictive policy, but also effective. Although a few employees have iPhones, they are personal. There was only one Taik iPhone at this school of over 100 people between 2008 and 2010.

The School of Design shows how respecting the lines of authority in the organization maintains its policies. The key thing was that any breach is done properly, respecting the lines of authority at the school, not by rushing to the shop and placing the deans in a situation in which they have to choose their side and risk losing face.

When we compare the School of Design to the first school, we see how the iPhone was just like any other symbol. The right to interpret what kind of symbol it becomes depends on whether the lines of power are respected or not. For designers, the iPhone is still not always a serious object. When this paper was written, one of the leading professors of the school said that he found it difficult to imagine how things could go wrong at the other school. In his own words, it is self-evident that if he buys toys, he uses his own money.

THE iPHONE IN A HACKER CULTURE

There was yet a third path to the iPhone. This path shows even better how power played a significant role in the domestication of this device. One of Taik's schools specializes in new media. iPhones started to appear in this school almost immediately after it was launched in the US in 2007, well before its launch in Finland in July 2008. Getting an iPhone did not become a big deal at this school, known for its technological orientation and hacker culture.

The first person to acquire an iPhone, in fact, was the then dean of the department, who bought the phone from California during research leave. When he returned to Helsinki, he hacked his phone so that it became operator independent, and simply put his Elisa SIM into the phone.

Several other members of the school followed suit. The iPhone was bought as a gadget, it was hacked, and it was used as a normal phone.

This became an unofficial policy of the school. The iPhone was defined as an internet gadget, and as a mini version of a laptop. As such, the phone policy was irrelevant, or not to be taken too seriously. Since the mission of the school is to explore new media technologies, the policy, the logic went, interfered with freedom of research and teaching. The same logic was applied to Apple: its products were not meant to be hacked, but if the possible rights of Apple became a hindrance to academic freedoms, the greater good justified hacking.

Hacking, of course, has been a cherished part of the culture of the school in talk and sometimes in practice. There is knowledge only hackers have, and if a school aims to know what is happening on the internet, it needs to support hacking as well. Of course, this was never an official policy; the dean was watching hacking through his fingers.

This was a temporary settlement. As soon as the iPhone became part of Elisa's selection, researchers and professors at the school switched to official products. The reasons were spelled out to me in an e-mail by a professor, who told me how he bought an iPhone from Cambridge, UK, in 2008, and used it as his personal phone with a Sonera SIM. He hacked the phone with Zibri's hack soon after returning to Helsinki to make it carrier independent. His reason was not to put a company card into it; he wanted to use it in Finland, which was not possible with a phone bought from the UK. This history was repeated soon after when he got a free iPhone 3G, which he was able to use in Finland, but not in New Zealand, where he went for a sabbatical. Again, he needed to hack it so that he could use his device with prepaid SIM cards in New Zealand and Australia. His hack was DevTeam's "yellow snow."

He switched to Elisa as soon as it became possible, however. The reason was that these hacks made any upgrade of iPhone's OS time-consuming and risky. When the iPhone 4 came to the market and a part of Elisa's selection, he wanted to get rid of the need to repeatedly hack his phone and risk losing his data. Also, as the baseband system in the iPhone 4 is unlocked, he explained, he can also use prepaid SIMs abroad. He writes:

> The down side to the hacked phone is upgrading to new releases of the iPhone OS. It's time consuming, boring and risky. As soon as Elisa started selling iPhone's to their business customers last year I upgraded

to the iPhone 4, which is a far superior device, especially re: camera and location technologies. I have no temptation to jail break this one as there are so many good, inexpensive apps available from the app-store and I don't have time for all the hacking upgrade nonsense. Also the baseband system (telephony and SIM related operating system) is already officially unlocked so I can use any prepaid sims I like abroad. (e-mail to the author, 31 March 2011: 18:03)

At this school, the iPhone was treated very differently from the two other schools. The reason for the existence of this school is new media and its tech-nologies. Not having experience of the iPhone would have been against the very charter of the school. For this reason alone, the process of getting iPhones was, if not legitimate, accepted. The acceptance started from the dean, pro-fessors and key researchers. After they had opened the path, they found little reason to say no to others who wanted to get an iPhone. Though everyone knew iPhones were against Taik policy, they easily found reasons not to care about this policy, as did the administration and the deans at other schools. The greater professional good became an unspoken policy.

THE iPHONE IS FREED

In 2010, the iPhone became freely available in two ways. First, Taik joined a larger university, named Aalto University. Officially, Taik is called Aalto University School of Art and Design, starting from 1 January 2010. Aalto's mobile phone policy aims at keeping the selection fairly small to ease upgrad-ing, but the iPhone is in its small selection. Second, as I have said, Taik's and Aalto's carrier of choice Elisa started to sell the iPhone in summer 2010.

After these changes, anyone entitled to a mobile phone at Aalto can purchase an iPhone, if his or her superiors accept the purchase. Installing an Elisa SIM into it is normal practice. It is also possible to install prepaid cards into these phones when traveling.

There is a twist, however. At Aalto University, the IT unit buys cheaper phones and takes care of "modest" phone bills. Specifically, this cheaper model is a desktop GSM or Nokia 2730, priced at 85 Euros, including value-added tax. Other models authorized by Aalto were Nokia E5 at 280 Euros, Nokia E7 at 590 Euros, Nokia N8 at 478 Euros, iPhone 16Gb at 598 Euros and iPhone 32Gb at 705 Euros. The policy also specified warranty: for Nokias, it was 24 months, for iPhones, 12. The following policy is on Aalto's intranet.

iPhones: Aalto moves from landlines to mobile phones in several stages[11]

In IT unit's development discussion 4.1.2011 (participants: the rector, vice rector, NN, MM, JJ, and KK), it was decided that the expenses for phone acquisitions are directly allocated to the units that buy the phones. The acquisition process and first use is administered by the IT unit. For those who have to change their landline phones to mobile phones due to changing the unit, the IT unit pays once for a decently priced mobile phone (Nokia 2730, desktop GSM). Aalto University acquires a connection through a centralized process administered by the IT unit, and pays modest phone bills for the connections.

This policy is valid immediately.

(Director, IT services Aalto University)

This policy creates an internal market at Aalto. Schools and other units have an incentive to push people to buy cheap models and use their phones sparingly. Specifically, it hits the more upmarket phones like the iPhone. The iPhone is more expensive than most other phones in Aalto's selection, and it is also more expensive to use. Price is a local matter: anyone interested in buying a phone has to get approval from his or her superior, who is typically the head of department. The superior needs to be convinced about the need to go for more expensive phones.

However, this policy leaks at many points. The difference between the price of the iPhone and more expensive Nokia models is small. In fact, the most expensive phones someone can buy at Aalto are Symbian phones, usually manufactured by Nokia. For example, Nokia Communicator has always been popular with senior professors and deans, and Aalto does not forbid these expensive models. Even though the website only suggests this, it is always possible to buy some of these very expensive models.

In terms of culture, two things come to play. First, price is related to organizational position. Several Nokia models like the communicator are far more expensive than iPhones. These upmarket products have been routinely bought and used by deans, senior professors, administrators and other frequent travelers who want to check e-mails wherever they are.

If history gives any clue to the future, the iPhone will finds its home in this category. Though the iPhone is not in the upmarket category alone, it is solidly there with the most expensive Symbians. In brief, purchasing an iPhone—or any other expensive phone—is a matter of whether the employee is able to justify it to his or her superiors. However, as always with justifications, some people need to justify more. For example, for the author of this chapter buying an iPhone was a matter of filling in an order form. Aalto's IT services bought him an iPhone 16Gb even though his department head forgot to authorize the purchase.

Second, there are different cultures of use. Designers have stayed in Symbian, and still generally distrust the iPhone at a technical level. It does drop calls and its sound quality is inferior to Symbian-based devices. Industrial designers, furthermore, have little interest in Apple's App Store. For media designers, on the other hand, the iPhone and its ecosystem are hugely interesting and important. Media designers have largely switched to the iPhone, although there is also a trend toward Android phones.

Obviously, it is too early to say much about these cultural matters. After all, the iPhone became available at Aalto in summer 2010, and the normal life cycle of a phone is two or three years. Few people have acquired new phones after the iPhone has been an approved option.

CODA: THE ROLE OF POWER

The tale told in this paper is not strange. It is not the first of its kind and certainly not the last. Similar battles have been fought over Apple and Microsoft cultures in computers, and many electronic devices like Walkmans, MP3 players, cameras and mobile phones have been objects of moral discussion at some point.[12] This chapter has argued that when gadgets like these enter formal organizations, they have to find their place not only morally, but also organizationally.

This chapter has looked at people in institutions: both those who run them and interpret their rules, and those who have little control to define what can be accepted. Specifically, this paper has looked at moral reasoning involved in making sense of whether it is possible to have an iPhone or not. The main inspirations were Everett C. Hughes and Charles Bosk, two Chicago sociologists who have pointed out that mistakes open a particularly interesting window to understanding organizations.[13]

The iPhone is a particularly interesting object for an organizational analysis for many reasons. As a phone, it is a tool, and obviously also a necessity at work. It was also the cool thing to have in 2007 and 2008. As several writers have noted, people have projected many kinds of meanings to mobile phones over the years, and the newest models have consistently been both desired and simultaneously frowned upon. The iPhone is no exception, although it is an Apple product, which has a long enjoyed cult following.

In fact, the iPhone was particularly suitable for raising these moral evaluations. Its slick industrial and smooth interaction design set the industry standard in 2007, especially in comparison to the market leader Nokia, whose phones all of a sudden looked dull in comparison. Against the Nokia-led Symbian background, the iPhone was a hot, dangerous object, and it was easy to turn it into an extension of self against the mass. As an object of passion, it was prone to clash with the standardizing mentality of bureaucratic rationality especially in a country in which most functions of the iPhone had been a commonplace for years.

The testbed for these ideas has been Taik, a large art and design school in Helsinki. Taik bought its phone services exclusively from Elisa Communications, which did not have the iPhone in its selection until 2010. This meant that anyone who wanted to have it had to find a way around Taik phone policy.

As we have seen, despite the policy, a few people managed to get an iPhone. However, there were major differences between the schools. At one school, the iPhone raised problems. There, people bought iPhones without asking for permission. Since the line of authority was not respected, the administration fought back and managed to restore the policy. In contrast, in media-oriented programs, the policy was permissive from the beginning. In the name of its mission, the dean and senior professors at one of the media programs authorized iPhone purchases and did not go after hacks. Their stance was a quiet acceptance, that is, "Don't ask, don't tell." In fact, at this program, the dean and senior professors set an example for such flexibility. In design schools, getting an iPhone required a good justification. Typically, this justification was research.

Although this chapter has shown that to understand mobile technologies, we sometimes need to place the locus of action within organizational dynamics rather than individual decisions, it also shows that there is nothing obvious about organizational policies. As the British critical management theorist Hugh Willmot has noted, rules are sense-making devices that provide people and organizations with ways to make sense of what is right and wrong.[14] Who has the power to use these rules to make sense of things and to forgive breaches is crucial in understanding how the iPhone was domesticated at Taik. Organizational rules and policies are not just flexible, they are also moral devices that management uses to draw a line between normal and abnormal, a mistake and a severe breach of morality.

In all, this chapter has taken an unusual angle on the iPhone. It has been treated as a moral object and a symbol that may endanger organizational policies. The chapter has shown how the iPhone threatened the established order at Taik, how knowing this established order led to many types of actions among Taik employees, and how administration and management was forced to form their own policies to deal with these actions. We have seen how these policies carefully balanced official policy with passion. Perhaps more than any other object recently, the iPhone became a matter of negotiation and also a display of power.

ACKNOWLEDGMENTS

I want to thank several colleagues for telling me about their policies regarding the iPhone and the reasoning behind those policies. A special thanks goes to Taik administration for telling me about its phone policy and about how the iPhone was handled in the organization. As former vice dean for the School of Design, I was one character in the story; my role in the story is that I authorized buying one iPhone for the project mentioned in section four of this paper.

NOTES

1. Timo Kopomaa, *The City in Your Pocket. Birth of the Mobile Information Society* (Helsinki: Gaudeamus, 2000).
2. For Sony Walkman, see Shuhei Hosokawa, "The Walkman effect," *Popular Music* 4, 1984: 165–180; Michael Bull, *Sounding Out the City* (Oxford: Berg, 2000); for generations of cameras, see Sarvas Risto and David M. Frohlich, *From Snapshots to Social Media. The Changing Picture of Domestic Photography* (London: Springer, 2011).
3. See also James E. Katz and Mark Aakhus, eds., *Perpetual Contact. Mobile Communication, Private Talk, Public Performance* (Cambridge: Cambridge University Press, 2002).
4. Rich Ling, "Fashion and vulgarity in the adoption of the mobile telephone among teens in Norway," in Leopoldina Fortunati, James E. Katz and Raimonda Riccini, eds., *Mediating the Human Body. Technology, Communication, and Fashion* (Mahwah, NJ: Lawrence Erlbaum, 2003).
5. Harold Garfinkel, *Studies in Ethnomethodology* (Englewood Cliffs, NJ: Prentice-Hall, 1967); Carolyn Baker, "Ticketing rules: categorization and moral ordering in a school staff meeting," in Stephan Hester and Peter Eglin, eds., *Culture in Action: Studies in Membership Categorization Analysis* (Washington, DC: University Press of America, 1997), 77–98; Ilpo Koskinen, *Managerial Evaluations at the Workplace* (Helsinki: Hakapaino, 1998).
6. Everett C. Hughes, *The Sociological Eye* (Chicago: University of Chicago Press, 1984).
7. Charles Bosk, *Forgive and Remember* (Chicago: University of Chicago Press, 1979).
8. As the critical management theorist Hugh Willmot has noted, organizational rules are sense-making resources. Hugh Willmot, "Studying managerial work: a critique and a proposal," *Journal of Management Studies* 24, 1987: 249–270; see also Mats Alvesson and Hugh Willmott, *Making Sense of Management: A Critical Introduction* (London: Sage, 1996). For seeing how these evaluations are done, see Koskinen, " *Managerial Evaluations at the Workplace.*"
9. See Roger Silverstone, Eric Hirsch and David Morley, "Information and communication technologies and the moral economy of the household," in Roger Silverstone and Eric Hirsch, eds., *Consuming Technologies* (London and New York: Routledge, 1999).
10. In this paper, I will follow local custom and refer to TeliaSonera as Sonera.
11. Translated by Koskinen. Names have been removed. Accessed 3 April, 2011.
12. Kopomaa, *The City in Your Pocket*; Hosokawa, "The Walkman Effect"; Bull, *Sounding Out the City*; Sarvas and Frohlich, *From Snapshots to Social Media.*
13. Hughes, *The Sociological Eye*; Charles Bosk, *Forgive and Remember.*
14. Willmot, "Studying managerial work."

Contributors

John Banks is a Postdoctoral Research Fellow in the Federation Fellowship program. His research interests focus on user-led innovation and consumer co-creation in participatory culture networks. Banks has a particular interest in videogames. From 2000 to 2005, Banks worked in the videogames industry for Brisbane-based Auran Games (www. auran.com) as an online community manager, focusing on the development of user-led content creation networks within the context of game development projects; he has published widely on research grounded in this industry background. Banks's current research continues to work at the interface of game developers and gamers as they negotiate emerging co-creation relations. Throughout 2007, he undertook ethnographic research with Auran Games on social network strategies for their massively multiplayer online game, *Fury* (http://www. unleashthefury.com).

Jean Burgess is a Senior Research Fellow in the Creative Industries Faculty and Deputy Director of the ARC Centre of Excellence for Creative Industries and Innovation, Queensland University of Technology (http:// cci.edu.au). She has published widely on issues of cultural participation in new media contexts and is the co-author of the first research monograph on *YouTube—YouTube: Online Video and Participatory Culture* (Polity Press, 2009), which has been translated into Portuguese (through Editora Aleph) and Italian (Editore EGEA).

Chris Chesher is a Digital Cultures academic at the University of Sydney who combines media studies and technology studies to interrogate the relationships between technological innovation and cultural change. His work has considered regimes of vision in console games, cinema and television; an actor network theory reading of the multiple uses of mobile phones at a U2 concert; how digital architectures promote archival breakdown; and the implications of software choices for knowledge mobility in Universities. Chesher's current research is looking at the aesthetics and politics of emerging robotics industries.

Kate Crawford is an Associate Professor and Deputy Director of the Journalism and Media Research Centre at the University of New South Wales, Sydney. She has published widely on social change and media technologies, particularly the terrain between humans, mobiles and social networks. Crawford has conducted research on mobile and social media use in Australia, India and the US.

Gerard Goggin is Professor of Media and Communications in the Department of Media and Communications, the University of Sydney. He is a leading figure in mobile media studies internationally. His books on mobiles have been extensively published by Routledge and include: *The Politics of Mobile Social Media* (forthcoming), *Mobile Technology and Place* (2012), *Global Mobile Media* (2011), *Mobile Telecommunications: From Telecommunications to Media* (2009), *Mobile Phone Cultures* (2008) and *Cell Phone Cultures* (2006). Goggin has also published books on the internet—*Internationalizing Internet Studies* (2009) and *Virtual Nation: The Internet in Australia* (2004)—and disability—*Disability and Media* (2012), *Disability in Australia* (2005) and *Digital Disability* (2003). Goggin holds a number of Australian Research Council grants for projects on: youth and mobile media; internet history in Asia-Pacific; and distribution of audiovisual material across all media platforms, including online and mobile.

Kay Gu is a Master candidate in the School of Media and Communication, RMIT University, Melbourne, Australia. She holds a Bachelor of Arts in Journalism from Fudan University (Shanghai, China) and worked for two years as a marketing executive. Gu's research focuses on new media, mobile communication and the impacts on Chinese young generation.

Larissa Hjorth is an artist, digital ethnographer and Associate Professor in the Games Programs, School of Media and Communication, RMIT University. Since 2000, Hjorth has been researching and publishing on gendered customizing of mobile communication, gaming and virtual communities in the Asia-Pacific—these studies are outlined in her book, *Mobile Media in the Asia-Pacific* (London, Routledge). Hjorth has published widely on the topic in national and international journals such as *Games and Culture journal, Convergence journal, Journal of Intercultural Studies, Continuum, ACCESS, Fibreculture* and *Southern Review* and in 2009 co-edited two Routledge anthologies, *Gaming Cultures and Place in the Asia–Pacific region* (with Dean Chan) and *Mobile technologies: from Telecommunication to Media* (with Gerard Goggin). In 2010 Hjorth released *Games & Gaming* (London: Berg).

Ilpo Koskinen was once a sociologist but has worked as a Professor of industrial design in Helsinki since 1999. His main research interests have been in mobile multimedia, the relationship of design and cities and interpretive methodology in design. Some of his main publications include *Mobile Image* (IT Press, Helsinki, 2002), *Empathic Design* (Helsinki: IT Press, 2003), *Mobile Multimedia in Action* (New Brunswick, NJ: Transaction Publishers, 2007) and *Design Research through Practice: From Lab, Field, and Showroom* (San Francisco: Morgan Kaufmann, 2011). Although he is not actively working on mobile media anymore, he still occasionally publishes a paper or two on this topic, if he thinks being one of the first contributors of the field gives him enough perspective to say something interesting.

Dong-Hoo Lee is Professor of the Department of Mass Communication at the University of Incheon, Korea. She obtained her PhD degree from the department of culture and communication at New York University. Her English language publications include articles in *The Information Society, Knowledge, Technology & Policy*, and the *Fibreculture Journal*, and chapters *in Mobile Technologies: From Telecommunication to Media* (2009) and *The Urban Communication Reader II* (2010). Her research interests are digital mobile communication, emerging visual media culture and medium theory.

Jack Linchuan Qiu is Associate Professor at the School of Journalism and Communication, the Chinese University of Hong Kong. He researches on information and communication technologies (ICTs), class, globalization and social change. His publications include *Working-Class Network Society* (MIT Press, 2009) and *Mobile Communication and Society: A Global Perspective* (MIT Press, 2006).

Daniel Palmer is a Senior Lecturer in the Theory Department of the Faculty of Art and Design at Monash University (Melbourne, Australia). He was formerly a Curator at the Centre for Contemporary Photography in Melbourne. His publications include *Twelve Australian Photo Artists* (2009), co-authored with Blair French, and *Photogenic: Essays/Photography/CCP 2000–2004* (2005). His writings on photography and media art have appeared in scholarly journals such as *Photographies, Reading Room, Philosophy of Photography, Angelaki* and *Transformations*.

Ingrid Richardson is Associate Professor in the Faculty of Arts, Education and Creative Media, and Director of the Centre for Everyday Life at Murdoch University, Western Australia. She has published widely on the cultural effects of new and emerging interfaces—including mobile media, the internet, games, urban screens, augmented reality, Web 2.0, social networking and remix culture—in national and international journals such as *Fibreculture, Continuum, Journal of Urban Technology, Mobilities* and the *Australian Journal of Communication*.

Nanna Verhoeff is Associate Professor at the Department for Media and Culture Studies at Utrecht University. Focusing on media in transition, she has written on early cinema in *The West in Early Cinema: After the Beginning* (Amsterdam University Press, 2006), and in her latest book *Mobile Screens: The Visual Regime of Navigation* (Amsterdam University Press, 2012), she analyzes mobility in screen-based media, ranging from panoramas, urban screens and handheld gadgets. Verhoeff's current project is a study on digital cartography, locative media and screen-based interfaces for digital (audiovisual) cultural collections.

Rowan Wilken is a Lecturer in media and communications at Swinburne University of Technology, Melbourne, Australia. His present research interests include domestic broadband consumption, digital technologies and culture, mobile and locative media, old and new media, and theories and practices of everyday life. He has published extensively on mobile media and is the author of *Teletechnologies, Place, and Community* (London: Routledge, 2011) and co-editor (with Gerard Goggin) of *Mobile Technology and Place* (London: Routledge, 2012).

Index